HUME'S MORAL
EPISTEMOLOGY

HUME'S MORAL EPISTEMOLOGY

Jonathan Harrison

CLARENDON PRESS · OXFORD
1976

Oxford University Press, Ely House, London W.1

GLASGOW   NEW YORK   TORONTO   MELBOURNE   WELLINGTON
CAPE TOWN   IBADAN   NAIROBI   DAR ES SALAAM   LUSAKA   ADDIS ABABA
DELHI   BOMBAY   CALCUTTA   MADRAS   KARACHI   DACCA
KUALA LUMPUR   SINGAPORE   HONG KONG   TOKYO

CASEBOUND ISBN 0 19 824566 1
PAPERBACK ISBN 0 19 875037 4

*Printed in Great Britain by
Butler & Tanner Ltd,
Frome and London*

to
VERA

PREFACE

M O S T of what Hume wrote about Moral Philosophy is to be found in Book III of *A Treatise of Human Nature* (published in 1740, a year later than the rest of the *Treatise*) and in *An Enquiry concerning the Principles of Morals* (published in 1751). In the following pages, I shall be concerned with only a small part of what he says in these two works. I shall be concerned only with those pages in which he tries to answer the question 'How do we know the difference between right and wrong, good and bad, virtue and vice?' These pages are to be found almost, though not quite, entirely in Book III, Part I, Sections 1 and 2 of the *Treatise*, and Section I and Appendix I of the *Enquiry*. There are, of course, a great number of other questions (sometimes more exciting ones) which Hume discusses in these two books, and on these I hope one day to write a commentary. But the problems which we shall consider here have both enough unity among themselves and are sufficiently independent of the rest of Hume's argument to be discussed separately. So far as I know, there is no accurate and detailed treatment of Hume's views on moral epistemology in the English language.

The main problem Hume devotes himself to is the question, which was a matter of considerable controversy among philosophers of his time, whether it was by reason or by sense that we determined the difference between right and wrong. If right and wrong were discovered by reason, then either the principles of morals are like the axioms of a mathematical system were supposed at that time to be—or they are like the theorems of such a system. If they were like the axioms, then a rational being was capable of apprehending, without argument, their necessary truth, and they were self-evident. If they were like the theorems, they could be deduced from such self-evident truths in a series of one or more logical steps. If, on the other hand, right and wrong were determined by sense, it becomes just a contingent matter of fact, which might be otherwise, what actions are right and wrong. What actions are right or wrong cannot in this case be determined by any amount of ratiocination; the question can only be settled by observation and experience. Hume, by and large, settled for the latter of these two views. In doing so, he supposed himself to be introducing 'the experimental method of reasoning into moral subjects'.

It is necessary to say a few words about Hume's terminology. More will be said about it later. He describes the view that right and wrong are determined by *a priori* ratiocination as holding that morality is determined by a comparison of ideas. He says this because he thinks that two and two are four—an *a priori* necessary truth, discoverable by reason—can be known to be true by comparing our idea of, say, two spots and another idea of two spots, and seeing that they must be equal in number to our idea of four spots. No appeal need be made to experience of what actually happens. The view that morality is an empirical matter, discovered by experience and observation, he describes as holding that morality is determined by our impressions. For example, no amount of *a priori* argument will tell us what colour grass is. In order to discover its colour, we must obtain an impression of grass—as it happens, by allowing light from the grass to reach our eyes—and it is only from such impressions that we can determine that it is green.

In the following pages I have summarized and tried to clarify Hume's arguments step by step, and subsequently commented on them. It did occur to me that the summaries might be omitted, and the reader simply referred to Hume's own words. However, I thought it would be inconvenient to have the argument and the comments in two separate volumes—mine and Hume's—and I did sometimes feel that my own words would be more easily intelligible to modern readers than were Hume's. However, I cannot over-emphasize the fact that my readers should constantly refer to what Hume himself actually says. Not to do this is not only to be unfair to a great philosopher, but foolishly to deprive oneself of the benefit which results from considering what he says with the respect and attention which his remarks deserve.

All references to the *Enquiry* are to the Selby-Bigge edition (revised by P. H. Nidditch, O.U.P. 1975), and will be prefixed by the letter E. All other references, except when it is explicitly stated otherwise, are to the *Treatise* (also edited by L. A. Selby-Bigge, O.U.P. 1888).

Chapter III may be omitted by those who have sufficient grasp of epistemology in general and Hume's epistemology in particular to understand what follows without reading it.

J.H.

CONTENTS

INTRODUCTION: HUME'S
STATEMENT OF THE PROBLEM

HUME's statement of the problem he is to attempt to answer is best put in his own words (456–7):

'It has been observ'd, that nothing is ever present to the mind but its perceptions; and that all the actions of seeing, hearing, judging, loving, hating, and thinking, fall under this denomination. The mind can never exert itself in any action, which we may not comprehend under the term of *perception*; and consequently that term is no less applicable to those judgements, by which we distinguish moral good and evil, than to every other operation of the mind. To approve of one character, to condemn another, are only so many different perceptions.

'Now as perceptions resolve themselves into two kinds, viz. *impressions* and *ideas*, this distinction gives rise to a question, with which we shall open up our present enquiry concerning morals, *Whether 'tis by means of our* ideas *or* impressions *we distinguish betwixt vice and virtue, and pronounce an action blameable or praiseworthy?* This will immediately cut off all loose discourses and declamations, and reduce us to something precise and exact on the present subject.

'Those who affirm that virtue is nothing but a conformity to reason; that there are eternal fitnesses and unfitnesses of things, which are the same to every rational being that considers them; that the immutable measures of right and wrong impose an obligation, not only on human creatures, but also on the Deity himself: All these systems concur in the opinion, that morality, like truth, is discern'd merely by ideas, and by their juxta-position and comparison. In order, therefore, to judge of these systems, we need only consider, whether it be possible, from reason alone, to distinguish betwixt moral good and evil, or whether there must concur some other principles to enable us to make that distinction.'

Comments

(1) 'Perception' is an odd word for the actions of the human mind. Loving and hating are clearly not in any normal sense perceptions. Furthermore, since we are aware in some way or other of many things which are not our perceptions, for example, the

contents of others' minds, and objects remote from us in space
and time, Hume is presupposing a distinction between immediate
and mediate awareness. He presumably thinks that the only things
we are *immediately* aware of ('present to the mind') are our own
perceptions. Whether he thinks we are immediately aware of *all*
our perceptions, he does not say. I believe that he did think this.

(2) The view that all perceptions are of two kinds, impressions
and ideas, is initially implausible, but there is not space to
comment on it. ('Ideas' are faint copies of impressions, as our mental image of a horse is a faint copy of a horse.)

(3) As we have seen in the Preface, Hume's question '*Whether
'tis by means of our* ideas *or* impressions *we distinguish betwixt vice
and virtue*' (456) is the question whether our knowledge of vice
and virtue is obtained *a priori*, by reason, or empirically, by
observation and experience. More will be said about this in
Chapter III.

(4) Though Hume suggests that he is going to show that
morality is not discerned by deductive reasoning, many of the
arguments which he later uses would, as we shall see, prove a more
extreme conclusion than this, viz. that morality is not discerned
by reasoning at all, whether deductive or inductive. And at least
one argument would, if valid, show that there were no moral
judgements to be established, whether by reasoning or in some
other way.

(5) The conclusion that morality is discerned by our impressions, not our ideas, is compatible with two very different views,
which Hume fails to distinguish. There is first of all the view that
moral judgements are not inferred from other propositions, which
we use as premises, but are known non-inferentially from our
impressions in some manner analogous to perceiving. For
example, we do not infer that grass is green; but our impressions
enable us to perceive that it is. The second of these two views
is that moral judgements are *inferred* from premises established
'by means of our impressions'. On both these two views impressions would be needed in order for us to distinguish between virtue
and vice. On the first view, impressions would enable us to know
what actions were virtuous in the way in which having impressions
enables us to know that this crow which I see before me is black.
On the second view, having impressions would enable us to know
what was virtuous in the way in which they enable us to know
that all crows are black. On the former view, no reasoning at all
is necessary in order for us to apprehend the morality of some-

thing. On the latter, reasoning *is* necessary, but is not sufficient, and is in any case not deductive reasoning, but inductive.

(6) If morality is discerned by deductive reasoning, then morality is 'eternal and immutable' in that the sentences which express the truths of morals, like the sentences which express the truths of mathematics or of logic, will, if they once express a truth, always express a truth. The sentence 'two plus two are four', provided that the words it contains do not change their meanings, cannot express something true at one time but false at another. The same will be true, if moral truths are discerned by deductive reasoning, of sentences such as 'promise-breaking is wrong'. But though, if moral distinctions are discerned by deductive reasoning, it follows that morality is immutable, if moral distinctions are not arrived at by deductive reasoning, it does not follow that morality is not immutable. More accurately, all that follows is that the sentences which at one time express moral truths *could*, at another time, express moral falsehoods. It does not follow that they do this. That gases expand when heated is a mutable truth in that gases *could* change in this respect, but not in the sense that sometimes they do. Perhaps it would be better to say that, if moral distinctions are not discerned by deductive reasoning, it follows that morality is immutable in one sense, but not in another. It follows that morality is not immutable, in that it follows that it would be logically possible for it to change, but it does not follow that morality is mutable in the sense that from time to time it does change.

(7) In a puzzling passage (456–7) Hume suggests that morality, unlike truth, is not discerned merely by the juxtaposition and comparison of ideas. This passage is puzzling for two reasons. Firstly it is not Hume's view that *all* truths are discerned merely by the juxta-position and comparison of ideas (and if it were Hume's view, it would be a silly one). Hume thought that it is only those truths which can be established by deductive reasoning which are established by the comparison of ideas; probable reasoning and perception enable us to establish truths in a different way. Secondly, Hume, in rejecting the view that morality, like truth, is discerned by ideas, is, intentionally or not, implying that apprehending moral distinctions is not apprehending truths. From this it would follow that there were no moral truths. We shall see that one of the arguments which he later uses would, if it were valid, establish that there were no moral truths. If there were no such things as moral truths it would follow *a fortiori* that moral truths were not 'deriv'd' from reason, as there would be no moral truth

to be discovered. This conclusion, however, is much more extreme than the view that moral truths are not established deductively. And it would, unfortunately, be incompatible with the view that Hume sometimes says is his own, that moral truths are discovered by a moral sense.

Let us now consider one by one the arguments by which Hume seeks to show that moral distinctions are not derived from reason, and that it is not by means of our ideas alone that we distinguish between vice and virtue.

I

HUME'S FIRST ARGUMENT: MORALITY MOVES US TO ACTION, WHILE REASON DOES NOT (457–8)

Hume's first premiss: reason cannot alone move us to action (413–7)

IN the *Treatise of Human Nature*, Book II, Part III, Section III, Hume supposes himself to have proved that, though almost all his predecessors thought that reason both could and ought to control the passions, in fact 'reason is, and ought only to be the slave of the passions, and can never pretend to any other office than to serve and obey them' (415). (By saying that reason *ought to be* the slave of the passions, Hume did not, inconsistently, imply that it could possibly be anything else; he simply meant that there was nothing to be deplored about reason's slavery.) Reason, according to Hume, is 'perfectly inert', and can never produce or oppose any action by itself (458), but only show the passions the way to satisfy themselves. By saying that 'reason alone can never be a motive to any action of the will' Hume seems to have meant that *reasoning* could not affect our behaviour, except in so far as it modified our beliefs, and that a *belief* could not move us to action, unless it was relevant to the satisfaction of some passion, desire, or need. The two ways in which Hume thought a belief could be relevant to the satisfaction of a passion were these: it might inform us of the existence of a desired object, for example apples in my neighbour's garden, or of the means to attaining this object, for example, begging, borrowing, or stealing them (460).

Hume, of course, never maintains anything so absurd as the proposition that my beliefs do not affect my behaviour; he merely holds that my beliefs *alone* do not affect my behaviour. Provided that I want to catch a train to London, the time at which I set out for the station will be determined by my belief concerning what time it leaves, but if I have no such desire, my belief by itself will not modify my behaviour in any way. Different actions can be produced by the same desires but different beliefs (as when both of us want to catch a train, but have different opinions about when it leaves), just as they can be produced by the same beliefs but different desires (as when both of us believe the tea is poisoned, but do not both want to die). Action is the product of belief and desire in conjunction, and can be changed by a change in either; but belief without desire is ineffective. On the question whether desire without belief is ineffective, Hume does not express an opinion.

When Hume says that reason alone cannot move us to action, I have interpreted him as meaning that beliefs alone cannot move us to action. Sometimes, however, he speaks as if it were not beliefs simply, but beliefs which were arrived at by means of reasoning, which could not alone move us to action. This is at least suggested, intentionally or not, by the statement that *reason* is the slave of the passions. However, it seems obvious that a belief moves us to action in precisely the same way, whether it is the product of reasoning or not; I may take arsenic because I believe it is sugar, whether reason entered into the formation of this belief or not. What Hume should have said is that beliefs move us to action only if they are relevant to the satisfaction of a passion, and that reasoning, whether demonstrative or probable, affects our actions only in so far as it produces beliefs which are so relevant. It is, however, a possibility which must not be lost sight of that Hume so far mistook the force of his own arguments as to suppose that he had shown only that beliefs which were the product of reasoning could not by themselves affect our behaviour, but allowed that beliefs which were not produced by reasoning could affect our behaviour in some other way. This possibility will be considered later (pp. 9–12).

That Hume was right in thinking that reason is the slave of the passions, in the sense that beliefs cannot alone move us to action, but simply inform us how desires can be satisfied, seems to me fairly obvious. Hume, however, adduces a proof of this fact. He thought that the fact that reason was the slave of the passions follows from the fact that actions or passions themselves cannot properly be described as reasonable or unreasonable (458). According to Hume, the only things which may be described as reasonable or unreasonable are 'ideas'. By 'ideas', in this context, he means beliefs. Ideas are copies of what they represent, or purport to represent, and so can be true copies, or false copies, in which latter case they are 'contradictory to truth and reason'. Actions and passions, on the other hand, are 'real existences', not copies, and have, unlike ideas, no reference to anything else. Actions and passions, therefore, cannot be true or false, conformable or contradictory to truth or reason. The only two ways in which a passion can be unreasonable is when it is founded upon a false or unreasonable belief, as when I fear something that I wrongly take to be dangerous, or when, in seeking to achieve its object, it acts upon a false or unreasonable belief about the means to securing this object (416). But 'where a passion is neither founded on false suppositions, nor chuses means insufficient for the end, the understanding can neither justify nor condemn it. 'Tis not contrary to reason to prefer the destruction of the whole world to the scratching of my finger. 'Tis not contrary to reason for me to chuse my total ruin, to prevent the least uneasiness to an *Indian* or person wholly unknown to me' (416).

Though Hume's conclusion, that beliefs alone cannot move us to action, but only point the way to satisfying our passions, seems to me to be fairly obviously true, the premiss from which he derives it, that 'reasonable' and 'unreasonable' can be applied only to beliefs, or to passions to the extent that they are based upon or directed by reasonable or unreasonable beliefs, seems to me to be false. My dictionary says that 'reasonable' can mean sensible, moderate, not absurd, fair. (There are other senses which, though they do not apply to beliefs, do not apply to passions either, but, usually, to people, for example, 'sound of judgement'.) It is not moderate, and is absurd, for me to prefer the destruction of the whole world to the scratching of my finger, or 'to chuse my total ruin, to prevent the least uneasiness to an *Indian* or person wholly unknown to me' (416), even if I do not have any false beliefs about the nature and consequences of what I am doing. It is certainly not sensible 'to prefer ... my own acknowledg'd lesser good to my greater' (416).

Of course, from the fact that a premiss is false and an alleged conclusion true, it does not at all follow that this premiss docs not imply the conclusion (though the premiss, since it is false, cannot be used to *prove* the conclusion). However, that only beliefs can be reasonable or unreasonable does not imply that reason is the slave of the passions. It could be, so far as I can see, that, even though only beliefs can be reasonable or unreasonable, I could nevertheless perform an action to which I was *not* impelled by any passion. Or rather, though I could not perform an action to which I was not impelled by any passion, this is not something which is rendered impossible by the fact that the epithets 'reasonable' and 'unreasonable' cannot be applied to passions.

Conversely, even if actions and passions could be described as reasonable or unreasonable, it would not follow that reason was *not* the slave of the passions. For, even if actions themselves could be reasonable or unreasonable, and I performed or omitted them for this reason, I would still need to be attracted by their reasonableness, or repelled by their unreasonableness. In other words, I would still need to have a passion for performing reasonable actions, if the fact that some actions were reasonable were not to leave me entirely indifferent to them. And even if some passions can properly be described as reasonable, it still would not follow that I possessed these reasonable passions, or that they were strong enough to move me to action in opposition to the other passions I possess. In any case, being moved to action by a reasonable passion would not be the same thing as being moved to action by a belief alone, without the co-operation of a passion.

Hume supposed that, if only beliefs could be reasonable or unreasonable, it followed that reason was the slave of the passions, for the following reason. 'Since a passion can never, in any sense, be call'd unreasonable, but when founded on a false supposition, or when it chuses means insufficient for the design'd end, 'tis impossible, that reason and passion can ever oppose each other, or dispute for the government of the will and actions. The moment we perceive the falsehood of any supposition, or the insufficiency of any means, our passions yield to our reason without any opposition. I may desire a fruit as of an excellent relish; but whenever you convince me of my mistake, my longing ceases. I may will the performance of certain actions as means of obtaining any desir'd good; but as my willing of these actions is only secondary, and founded on the supposition, that they are causes of the propos'd effect; as soon as I discover the falshood of that supposition, they must become indifferent to me' (416–7). It is not that I disagree with what Hume says in this passage; in fact, I agree with it. All Hume proves here, however, is that *given* (a) that actions themselves are not capable of being reasonable or unreasonable, and (b) that we desire things having a given feature, then what we do will be determined by our beliefs about what things possess this feature, and how they are to be brought about. It does not follow, though I think it is true, that our beliefs will not move us to action, unless they are beliefs about what things possess some desired feature, and how these things are to be secured.

As we have seen, Hume's premiss, that only beliefs are capable of being reasonable or unreasonable, is false. However, I do not think it matters very much that it is false, and this for the following reasons. For one thing, as we have seen, Hume's conclusion, that beliefs alone cannot modify behaviour, and need the co-operation of a passion, is true. Secondly, in the senses of 'reasonable' and 'unreasonable' in which these words apply to actions and passion, it would be a mistake to jump to the conclusion that it is reason that discovers what actions or passions are reasonable or otherwise. For example, 'reasonable' can mean 'fair', and, though it *may* be our *reason* which discovers what is *reasonable* in the sense of fair, it would be a mistake to take it for granted, just because the words are similar, that this must be so. Thirdly, if the claim that reason ought to be the master of the passions, whether it necessarily has to be or not, means not that we ought to perform actions to which we are not prompted by any passion, which is impossible, but that we *ought to perform* actions which are reasonable, or be actuated by reasonable passions, then this claim, which is a substantive moral one, is not obviously a claim which is discovered by reason. In any case, it is a claim which is very probably false. The claim that we ought to be reasonable, in whatever sense, suggests that we should always keep our behaviour within bounds, be moderate, compromise, adopt the middle course. This, no doubt, is as a general rule highly commendable, but one cannot but suspect that some of the best, as well as some of the worst, human actions go beyond the bounds set by what is reasonable. It would be a pale and inaccurate description of the action of a man who throws himself on a bomb to protect with his body the lives of his friends that it was, in the circumstances, the reasonable thing for him to do.

Hume's second premiss: knowledge of morality can move us to action (457)

Hume next argues that, though reason alone has no influence on our passions and actions, morality has. This is suggested by the fact that it is a branch of practical, rather than speculative philosophy, and confirmed by common experience, which 'informs us, that men are often govern'd by their duties, and are deter'd from some actions by the opinion of injustice, and impell'd to others by that of obligation' (457).

It is possible, for reasons which I shall explain later, to interpret Hume's second premiss in two ways. According to the first interpretation, he is saying that moral beliefs alone influence our actions and passions. According to the second, he is saying that *so-called* moral beliefs, or what purport to be moral beliefs, alone move us to action, but what appear to be moral beliefs are not in fact beliefs at all. The language he himself uses suggests the former of these two interpretations, but in some ways the latter fits in better with his argument, as we shall presently see. That there were no genuine moral beliefs, and so no true moral beliefs, is also suggested, as we have seen (p. 3) by his remark that morality, unlike truth, is not discerned merely by the juxta-position and comparison of ideas.

Hume's conclusion: morality not discovered by reason (457)

I have suggested that two interpretations of Hume's first premiss are possible, firstly, the more extreme interpretation, that beliefs do not alone move us to action, and secondly, the less extreme (and also less plausible) interpretation that beliefs which are arrived at by a process of reasoning do not move us to action. Clearly what Hume's argument proves, even if we grant him his second premiss, depends very much on which of these two interpretations we adopt. If we adopt the second, we get the argument: beliefs which are arrived at by a process of reasoning do not alone move us to action; moral beliefs do alone move us to action; hence moral beliefs are not arrived at by a process of reasoning. If we adopt the first interpretation, we get the argument, proving a much more drastic conclusion: beliefs do not alone move us to action, so-called moral beliefs do alone move us to action; therefore what are alleged to be moral beliefs are not beliefs at all. (If we adopt the first interpretation of the first premiss, we have to say 'so-called moral beliefs' instead of 'moral beliefs'; for otherwise Hume's second premiss, that moral beliefs alone move us to action, is just *incompatible* with his first premiss, that beliefs alone do not move us to action.)

Comments

(1) As we have seen, the chief, if not, strictly speaking, the only argument Hume brings to prove his first premiss, that reason is the slave of the passions, is that 'actions of the will' are not properly described as being either reasonable or unreasonable (458). This is best treated of when we later come to consider Hume's second argument to show that moral distinctions are not derived from reason (pp. 17–26).

(2) Hume's first premiss, interpreted as meaning that beliefs alone do not move us to action, does seem highly plausible; indeed, I think it is true. *Why* we should think it true is another matter. Have we learnt, by observation and experience, that there has never been an observed case of an action to which we have not been moved by a passion, and concluded that there never is, has, or will be? Or is it evident, on reflection, that every action is produced by a passion, as it is evident, on reflection, that every effect is preceded by a cause, or that there cannot be a mountain without a valley? If it is, is it possible to show why it is thus evident?

(3) Hume seems to me to be wrong when he says that an action can be unreasonable only if we are misinformed about the existence of a desired object, or of the means of attaining it (416). Obviously, our beliefs about how what we want is to be got will affect our behaviour. But there is no reason why the object of a passion must exist, if that passion is to be satisfied; the apples I want *may* exist in my neighbour's garden, but, on the other hand, I may have to grow them. I would have been better if Hume had said that reason could move us to action by causing us to believe that some state of affairs, for example, our possession or consuming apples, would, if it existed, have a certain feature or features which would cause it to attract us, and by showing us how this state of affairs could be brought about. Not all desires, incidentally, are desires that a future state of affairs be realized. We might, for example, have a strong desire that Richard III be not responsible for the murder of the princes in the Tower.

(4) Hume seems somewhat confused about what his second premiss needs to be, in order to entail a conclusion. His first premiss, be it remembered, was not that reason does not move us to action, which is obviously false, but that reason *alone* does not move us to action, i.e. does not move us to action without the co-operation of a passion. In order to show that, where morality is concerned, the case is different, he needs to prove that morality, unlike reason *alone*, moves us to action (i.e. moves us to action *without* the co-operation of any passion). However, all he is entitled to regard as obviously true, all he adduces any reason for, is that morality moves us to action in some way or other, and this, he himself concedes, reason does, too. The proposition he needs to prove a conclusion—that moral beliefs *alone* move us to action—and the proposition he adduces evidence for—that moral beliefs *do* move us to action—are totally different.

(5) The first of the two possible interpretations of Hume's conclusion mentioned earlier, that beliefs which are arrived at by a
process of reasoning do not move us to action, fits in very well
with the moral sense theory which Hume alleges he holds. According to a moral sense theory, we do not infer actions to be virtuous,
any more than we infer that a rose we can see is red; just as we
see that the rose is red, we 'see' that certain actions are virtuous,
by means of our moral sense. Hence, according to a moral sense
theory, moral beliefs are not arrived at by a process of reasoning
and so they could alone move us to action even if beliefs which
are the result of reasoning do not. Unfortunately, however, it is
most implausible to suggest that how a belief is arrived at makes
any difference to whether or not it moves us to action. I may see
that the gun is loaded, or infer that it is, but, in either event, I
shall be equally reluctant to pull the trigger, given that I do not
want to kill anybody.

(6) This interpretation of Hume's conclusion would also fit in
well with what he himself suggests that he is going to do in the
puzzling passage (456–7) already referred to (p. 3): 'Those who
affirm that virtue is nothing but a conformity to reason; that there
are eternal fitnesses and unfitnesses of things, which are the same
to every rational being that considers them; that the immutable
measures of right and wrong impose an obligation, not only on
human creatures, but also on the Deity himself; All these systems
concur in the opinion, that morality, like truth, is discern'd merely
by ideas, and by their juxta-position and comparison.' (456–7.)
In rejecting the view that reason discovers morality by the juxta-
position and comparison of ideas, Hume should mean that he is
rejecting the view that it is discovered by deductive or demonstrative reason, for it is, according to Hume, deductive or demonstrative reason that operates by the juxta-position and comparison of
ideas (69–73). There are, however, even according to Hume, many
true beliefs which are not arrived at by demonstrative reasoning.

(7) The second possible interpretation of Hume's conclusion,
that so-called moral judgements or opinions are not really judgements or opinions at all, is the one suggested by the most plausible
interpretation of his first premiss, which is that *beliefs* alone do
not move us to action. Hume's argument then becomes: 'Beliefs
alone does not move us to action; apprehending a moral distinction
does alone move us to action; hence apprehending a moral distinction is not a matter of having a belief.' This interpretation is
also suggested by the passage in which he says that morality is

not 'like truth ... discern'd merely by ideas, and by their juxta-
position and comparison' (456–7), and it is the view which those
who hold the once fashionable opinion that there are no moral
judgements or propositions, and that the function of ethical sentences is not to convey true or false information, but to express
favourable or unfavourable attitudes, or to command, exhort, or
persuade, would naturally like to be his, to the extent that they
value his support for their view.

However, there is no evidence that Hume ever even conceived
this possibility, other than that it would fit in very well with *some*
of the things he says, if he held it. In any case, it is very difficult
to resist the conclusion that there *are* moral judgements or propositions, when one considers how closely ethical sentences
resemble sentences which everybody agrees do express judgements or propositions. That the cat is in the larder is agreed by
everybody to be a statement, proposition, or judgement; one can
think, believe, or be convinced that the cat is in the larder; one
can be sure the cat is not in the larder; one can wonder whether
the cat is in the larder or not; one can agree with Smith that the
cat is in the larder, and disagree with Jones, who maintains that
it is not. Similarly, one can think, believe, or be convinced that
it is one's duty to resign; one can be sure that it is not one's duty
to resign; one can wonder whether it is one's duty to resign; one
can agree with Smith that it is not, and disagree with Jones, who
thinks it is. Indeed, if you *define* a proposition as that which it
is possible to believe, disbelieve, deny, doubt, argue about, then
that it is my duty to resign must be a proposition, for one certainly
can believe it, disbelieve it, deny it, and so on. Perhaps, then, those
who attribute the non-propositional theory to Hume are not doing
him the service they imagine. Incidentally, the view that there are
no moral judgements or propositions is incompatible with many
of Hume's own positive contentions about the nature of our knowledge of morality, including his contention that moral distinctions
are derived from a moral sense.

(8) *Prima facie*, it does seem as if Hume is making a quite unnecessary fuss about the way in which our knowledge of morality
moves us to action. The rationalist (and anyone else who thinks
that there are such things as moral judgements) can say—and it
seems perfectly obvious that he would be right to say—that there
is such a thing as a desire to do what is right, to behave as one
ought, or to do one's duty. Given that there is such a desire, then
my beliefs about morality may affect my behaviour in precisely

the same way as any other beliefs affect my behaviour, that is, by being relevant to the satisfaction of a passion, in this case, my passion for morality. Clearly, if I want to do what is right, then my beliefs about what is right will affect my behaviour, just as, if I want to catch a train, my beliefs about what time it leaves will affect my behaviour. On the other hand, if I have no desire to act rightly, then my beliefs about morality will not affect my behaviour. Indeed, it does seem quite natural for anyone without an axe to grind to draw a distinction between *wanting* to behave properly and *believing* that a certain course of action is the proper one to take. I may know (or think I know) what my duty is, but not want to do it; on the other hand, I may want to do my duty, but not know what it is. Hume himself draws this distinction on page 465 of the *Treatise* when he says ''Tis one thing to know virtue, and another to conform the will to it.' But if there is such a thing as a desire to do what is right, then it is quite easy for the rationalist to explain how an apprehension of moral distinctions affects my behaviour, and Hume's argument is left without any semblance of plausibility whatsoever.

(9) This is not, of course, to say that many rationalist Philosophers have not held that reason by itself both could and ought to govern the passions. Plato, for example, in the *Republic* (Book IV, 439) argues that because, though thirsty, I may nevertheless be deterred from drinking water which I believe to be poisoned, there must be an element in the soul (reason) which controls passions such as thirst. It never seems to have occurred to him that I refrain from drinking only because I am moved by another passion, the passion to stay alive. Hence my not drinking the poisoned water is just another case of one passion prevailing over another. Kant (writing after Hume) not only thought that it was reason which gives us the moral law, but also that knowledge of the moral law must sometimes itself move us to action; otherwise we are acting merely from inclination (in Hume's language, from passion), and our action will be morally worthless. Indeed, the view that reason both could and should control our passions has imbued literature and common thought for centuries. Hume's criticism of this rationalist view goes home, but it must not be forgotten that his main target was not the view that reason controls the passions, but the view that reason apprehends moral distinctions, and that his chief end in rejecting the former view was that he thought that, if you did reject it, you must reject the latter view also. In thinking this, he was totally mistaken.

(10) Hume not only refutes the view that reason alone can move us to action; he bolsters up this refutation with an explanation of how it comes about that this erroneous view could have been supposed to be true. The reason, he thinks, is that there are certain calm passions, benevolence, self-love, and love of virtue, which, just because they are calm, affect the human mind but little. Consequently their presence is overlooked, and actions, which are in fact due to them, are wrongly supposed to be the product of reason (417). The expression 'calm passion' would not sound as odd to eighteenth-century ears as it does to ours, for then the word 'passion' did not have its modern overtones of violence. For the same reason Hume, in saying that reason ought to be the slave of the passions, was not recommending a life dominated by violence and lust. The passions which he thought ought to prevail were the calm passions already mentioned, benevolence, self-love, and love of virtue.

(11) It must not be supposed that this criticism of Hume's first argument against rationalism is of historical and antiquarian interest only. It is very commonly supposed, even by modern philosophers, that our knowledge of morality is practical in some way in which our knowledge of, say, chemistry is not, and from this it is concluded that the outcome of moral reasoning cannot be a piece of knowledge just like the outcome of historical or scientific or mathematical reasoning. This latter would be something theoretical, while our knowledge of morality must answer practical questions, must be 'action guiding', and must tell us, not what is the case, but what to do. Though not the only argument purporting to prove this conclusion, one of them certainly is this, that unless there is some radical difference between our knowledge of morality and so-called 'theoretical' knowledge, it is impossible to explain how our knowledge of the former can guide our behaviour. There is not, however, as the above criticism of Hume has shown, the least difficulty. Given that we want to do what is right, the 'practical' knowledge that a given action would, if performed, be right, could move us to action in precisely the same way that, given a desire for sweet things, the 'theoretical' knowledge that sugar is sweet can move us to action. Indeed, it is so easy to explain how our knowledge of morality can both be theoretical and move us to action, that it is difficult to see why so much weight has been attached to Hume's argument, either by Hume or by like-minded successors.

II

HUME'S SECOND ARGUMENT: 'REASONABLE' AND 'UNREASONABLE' CANNOT BE APPLIED TO ACTIONS (458–63)

Hume's first premiss: 'reasonable' and 'unreasonable' apply only to beliefs (415, 458)

THE first premiss of Hume's second argument against the view that moral distinctions are discerned by reason is that the epithets 'reasonable' and 'unreasonable' can be applied to ideas (or beliefs) only, and never to the real existences of which these are true or false copies (415, 458). This proposition has already been discussed (pp. 6–9). My belief that there is a cheque in the envelope can be true or false, reasonable or unreasonable, but the cheque itself, which is a 'real existence', cannot properly be said to be either. Beliefs, which are copies of things, can be true or false copies, and be reasonably or unreasonably formed, but the things of which they are copies are 'compleat in themselves', have no reference to anything else, and so cannot be true or false, reasonable or unreasonable (458).

Hume's second premiss: actions are 'real existences', not beliefs (458)

Hume's second premiss is that 'our passions, volitions, and actions' belong to the class of real existences, not to the class of copies; hence they are complete in themselves, and cannot be said to be true or false, reasonable or unreasonable (458). In more modern language, as we have seen, they are not beliefs or propositions, and so cannot without impropriety be described as reasonable or unreasonable, true or false. This is why it is 'not contrary to reason to prefer the destruction of the whole world to the scratching of my finger' (416). By this Hume does not mean that such behaviour accords with reason, but that this act of choice, being a real existence, not a copy, can be described neither as conforming to reason nor as being contrary to it.

Hume's conclusion: the morality of actions not discerned by reason (458)

Hume concludes *a fortiori* that, since actions cannot be described as true or false, reasonable or unreasonable, they cannot *derive* their merit or their demerit from their conforming to reason (i.e. their being reasonable) or their failing to conform to reason (i.e. their being unreasonable) (458). Still less, presumably, could their being praiseworthy or blameworthy simply *consist* in, or be identical with, their being reasonable or unreasonable. And Hume, as we shall see, seems to think that this implies, or is the same thing as (pp. 24–6), that reason does not apprehend the fact that actions are moral or immoral.

Hume's 'indirect' proof of his conclusion (458)

Hume says that not only does the above argument prove *directly* the conclusion that actions do not derive their merit from their conformity to reason, it also proves the same thing *indirectly*, by showing us that 'as reason can never immediately prevent or produce any action by contradicting or approving of it, it cannot be the source of moral good and evil, which are found to have that influence' (458). By saying that this proof is 'indirect', he means that not only does his argument prove that moral distinctions are not derived from reason, in the manner already explained, it also proves that reason cannot alone move us to action (which is just the first of the two premisses of Hume's first argument). The direct proof is that since true and false, reasonable and unreasonable, apply to ideas, copies, or beliefs, and not to actions, reasonable and unreasonable cannot be the same as moral and immoral; hence moral distinctions cannot be derived from reason, which concerns itself only with what is true or reasonable. The indirect proof is that since true and false, reasonable and unreasonable, apply to ideas, copies, or beliefs, and not to actions, they are not the same as moral and immoral, from which it follows that reason cannot be concerned about them, and so cannot alone move us to action, which is one of the premisses of Hume's first argument. From this, with the addition of the other premiss, that apprehending a moral distinction *can* alone move us to action, it again follows that moral distinctions are not derived from reason.

Two 'abusive' senses in which actions and passions may be
reasonable or unreasonable; firstly, they may be
caused by reasonable or unreasonable beliefs about
(a) matters of fact (459–60)

Hume next considers the possibility that, though actions and passions cannot properly be said to be contrary to reason, they can be so described in a wide or improper sense, if they are either *caused by* judgements which are contrary to reason, or themselves *cause* such judgements (459).

If my action is caused by a false or unreasonable judgement, this judgement must be either one of fact (459), or one of right (460). If I act on a false or unreasonable judgement on *a matter of fact*, this must be either a mistake about the means to attaining some end, which I desire, or a mistake about the pleasure I would get from this end, if I succeeded in realizing it (459). Hume argues that, though actions (and, presumably, volitions and passions) can be, in an extended sense, unreasonable, in that they are based upon an unreasonable judgement, their being so based cannot be grounds for saying that they are immoral. For one thing, mistakes about matters of fact are often unavoidable, and so innocent (459). (Hume himself never asserts, though in one place he suggests, that mistakes of fact are always innocent, nor does his argument need this assertion.) For another thing, since the agreement or disagreement with reality of the judgements upon which our actions are founded does not admit of degrees, all virtues and vices would be equal (460).

Secondly, they may be caused by reasonable or unreasonable
beliefs about (b) matters of right (460)

Hume argues that the immorality of an action cannot consist in its being unreasonable in the sense that it is based upon a mistake of *right*, for ''tis impossible such a mistake can ever be the original source of immorality, since it supposes a real right and wrong; that is, a real distinction in morals, independent of these judgments'.

The second 'abusive' sense: actions may be reasonable
or unreasonable in that they cause reasonable
or unreasonable beliefs (461–3)

Hume asserts that 'a late author, who has had the good fortune to obtain some reputation', has argued that an action derives its immorality from having false or unreasonable judgements as its *effects*. ''Tis certain,' says Hume, 'that an action, on many occasions, may give rise to false conclusions in others; and that a person, who thro' a window sees any lewd behaviour of mine with my neighbour's wife, may be so simple as to imagine that she is certainly my own' (461). But, Hume argues, the false effects of my action are 'accidental', though this is questionable, in the example he himself gives. The final refutation, however, comes in a footnote: '... if I had used the precaution of shutting the windows, while I indulg'd myself in those liberties with my neighbour's wife, I should have been guilty of no immorality; and that because my action, being perfectly conceal'd, wou'd have had no tendency to produce any false conclusion' (461 n.).

(The word 'judgments', on p. 463, l. 1 of the Selby-Bigge edition of Hume's *Treatise*, is presumably a misprint for 'actions'.)

Comments

(1) Throughout this argument Hume fails to draw any distinction between the rationality of a belief and its truth. Obviously, some very irrational beliefs, such as that ill will befall me on Friday 13th, sometimes turn out to be true, and some extremely rational beliefs, such as that I will not inherit a fortune, sometimes turn out to be false. However, I do not think that drawing a distinction between rationality and truth would, or ought to have, made Hume wish to alter his argument in any way. His argument needs the premiss that 'true' and 'reasonable' apply to ideas or beliefs, but not to persons or actions, and that, to the extent that it is true at all, is equally true of 'true' and of 'reasonable'.

(2) A more serious difficulty with Hume's argument is that one of its premisses is not, strictly speaking, true. Besides speaking of reasonable and unreasonable beliefs, we speak of reasonable or unreasonable people, reasonable or unreasonable behaviour, actions, impulses, policies, demands, or requests. Besides speaking of true or false beliefs, we speak of a true friend, a true die, the true heir, a false knave, a false shot, a false start, a false move, and so on. Not only do we speak of contradictory beliefs, we also speak of contradictory behaviour. To get over this difficulty, presumably, Hume would have to contend that the predicates true and false, reasonable and unreasonable, are applied to things other than beliefs in different senses from that in which they are applied to beliefs, and perhaps that, in these senses, too, they cannot con-

stitute morality and immorality. This is, indeed, what he himself does for 'unreasonable' when he considers the possibility that it means 'caused by, or based on, a false belief', or 'resulting in a false belief'; doubtless, he could use similar arguments to show that being false and unreasonable, in any other of the senses I have mentioned above, cannot constitute immorality. A false move, for example, is not the same thing as an immoral move. Hume is obviously inclined to think that, if the epithets 'false' and 'unreasonable' apply to things other than beliefs at all, then they can be reduced to the sense in which they apply to beliefs; an unreasonable action, he suggests, may be an action based on an unreasonable belief (415–16). I do not think, however, that it is always the case that 'false', 'unreasonable', etc., as they apply to things other than beliefs, can be defined in terms of the sense of these words in which they apply to beliefs. A false shot, for example, is the result of clumsiness or nervousness, not of mistaken or ill-founded belief, and a woman may be unreasonable not because she cannot reason, or because she believes what is false, but because she is in the grip of impulses which are violent, inconsistent, and which she cannot or will not control (She may even have a reasonable belief that unreasonable behaviour will increase her chances of getting her own way.)

(3) Hume's 'indirect' proof of the proposition that the immorality of actions cannot derive from their being contrary to reason (458) is not as perspicuous as it might be. He thought that the fact that actions could not themselves be contrary to reason showed that they could not derive their immorality from their being contrary to reason; this is his direct proof. He also thought that the fact that actions could not themselves be contrary to reason showed that reason could not alone move us to action, from which he thought it followed that reason could not apprehend moral distinctions; this is his indirect proof. But *why* did he think that the fact that actions could not themselves conform or fail to conform to reason showed that reason alone could not move us to action? I think the answer—if I may be allowed to fill out Hume's argument a little—is that he supposed that if reason could, *per impossibile*, be a moving force, the only thing about any contemplated action which could attract or repel it would be the reasonableness or unreasonableness of this action; hence, since behaviour was incapable of being reasonable or unreasonable, *nothing* about an action could appeal or fail to appeal to reason. However, as we have seen, even if actions could be reasonable or

unreasonable, reason alone would still be incapable of moving us to action; a passion for performing reasonable actions would still be needed. Propositions can certainly be reasonable or unreasonable, true or false, but this fact would not be of any interest to us, were it not for a passion for truth and rationality, either for its own sake, or because of the very considerable heuristic advantages of having true and reasonable beliefs.

(4) I see little to cavil at in Hume's contention that, even if actions can be called contrary to reason, in a loose sense, by being based upon a false or unreasonable belief about a *matter of fact*, this cannot be the reason for their being immoral. It is true that in one place (460) he suggests, wrongly, that mistakes of fact are always innocent—whereas, in fact, such mistakes are often the result of character-failings such as carelessness, negligence, laziness, and conceit—but, in the passage (460) where he makes this suggestion ('Shou'd it be pretended, that tho' a mistake of *fact* be not criminal, yet a mistake of *right* often is'), he may simply be putting forward the view of an imaginary objector, not his own view. His argument, nevertheless, only needs the certainly true premiss that mistakes of fact are often innocent (which is what he himself says on page 459) whereas, according to the view Hume is refuting (that a wrong action is one based on a mistake about a matter of fact), actions based on such mistakes would be *always* guilty, whether the mistake was avoidable or not. His contention that the view he is arguing against also leads to the absurd conclusion that there are no degrees of viciousness, since falsehood does not admit of degrees, is less convincing (460). Not all mistakes are equally serious, and not all false assertions are equally far from the truth, and I suppose it could be argued, though not with plausibility, that the immorality of an action is greater, as the mistake on which the action is based is more serious. By saying 'nor will there be any difference, whether the question be concerning an apple or a kingdom' (460), Hume presumably means that another unacceptable consequence of the view which derives the immorality of an action from its being based on a mistake of fact is that it makes no difference to the virtue of an action whether the mistake in question is on an important or an unimportant matter. Whether or not it does make such a difference I do not know. (Wollaston held that 'the degrees of evil or guilt are as the *importance* and *number* of the truths violated'. See *The Religion of Nature Delineated*, in Vol. I of *British Moralists 1650–1800*, edited by D. D. Raphael, 1969, p. 251.) Certainly we ought to take more

care not to make a mistake of fact when the fact might be that we do a serious, as opposed to a minor, wrong.

(5) Hume now passes to actions which he thinks might loosely be called irrational in that they are based on a mistake of right. He says that such mistakes are sometimes criminal (460) (which does suggest that he thought that mistakes of fact were never criminal, or there would be a difference between mistakes of fact and mistakes of right, which he does not say are *always* criminal) and so the actions based on them can be immoral. Nevertheless, being based on such a mistake cannot be the only reason why any action is ever immoral, for this mistake must itself be a mistake about morality; given that there is a (primary) morality, I can make mistakes about it, and my making these mistakes can give rise to a secondary form of morality, but the secondary form of morality could not be the only sort, or there would be nothing about which I could be mistaken. It is rather as if Hume were arguing that, though hypochondria could be an illness, it must be a derivative one, i.e. the illness of supposing oneself to have some *other* illness; hence it cannot be the only illness; there must also be the other illnesses which one wrongly supposes oneself to have.

However, either the analogy with hypochondria is false, or Hume's argument breaks down. Though everyone except Christian Scientists knows that it is not, hypochondria *could* be the only illness, i.e. the only illness anyone ever actually has, though there must be names for other illness, illness which one might have, and wrongly supposes one does have. Similarly there might be names for other possible vices, such as laziness, which one wrongly supposes one has, though the only real vice is that of indulging in a way of life which one wrongly supposes to be idle (or of indulging in an idle life which one wrongly supposes to be on that account vicious).

However, the analogy between the view Hume is attacking and the view that the only illness men ever suffer from is hypochondria is false. A truer analogy would be with a view which held, not that hypochondria was the only actual illness, but that it was the only conceivable illness, that illness and hypochondria were one and the same thing. Such a view would surely fall to arguments analogous to Hume's. There must be illnesses other than hypochondria, even if no one ever has them; or there could be no such thing as hypochondria.

The reason why an objection, which seemed conclusive against the view that action based on a mistake about morality is the only form of immorality, was not conclusive against the view that hypochondria is the only illness anyone ever actually has, is this. The disease a man who thinks he is ill has got is not the disease he wrongly thinks he has got—this would be contradictory—but some other disease; he thinks wrongly that he has cancer, but he has got hypochondria. But the defect of the man who thinks he is vicious is the same defect as the defect he wrongly thinks he has got, i.e. vice. Hence, contradictions arise. The man who rightly thinks his action is right is acting rightly on two counts, because his action is right, and because he is right about its being right; and the man who wrongly thinks it is right is acting wrongly on two counts, because if he is wrong in thinking it right, it must be wrong, and because, since he is mistaken in his moral assessment, he must be acting wrongly. But the man who rightly thinks his action is wrong is acting rightly on one count—for if he is not acting on a mistake, he is acting rightly—but, on another count, he is acting wrongly, for if he is right in thinking it wrong, he must be acting wrongly. The man who wrongly thinks he is acting wrongly is also acting rightly on one count, and wrongly on another—wrongly because he is making a mistake, rightly because he is mistaken in thinking his action wrong. It is to avoid difficulties such as these, though less severe, that those who wish to hold that it is in fact immoral to act in a way which one wrongly supposes to be immoral have been forced, implausibly, to distinguish between two senses of 'right' and 'wrong'.

It should not be overlooked that Hume, all through the paragraphs we are at the moment discussing, talks as if there were moral beliefs, and as if a mistake of right were a false belief. If the object of the first argument were to prove that morality, since it moves us to action without the co-operation of any passion, cannot be a question of having beliefs, these passages would be plainly inconsistent with it.

(6) I wish to end with one final comment, a criticism which is, I believe, quite fatal to this second of Hume's objections to rationalism in ethics. All this consideration of whether Hume's premiss, that actions, as opposed to beliefs, cannot be contrary to reason, is true, and whether his conclusion, that actions cannot derive their immorality from their being contrary to reason, follows from it, is beside the point, for Hume is guilty of *ignoratio elenchi*. The position he refutes is not the position his antagonists maintain, or, at any rate, need to maintain, nor is it the one that he himself earlier says he is going to demolish. The position he

says he is going to demolish is that it is reason which discovers whether an action is immoral or moral (455); the position he does, to a very large extent, demolish is the one that rationality and morality, irrationality and immorality, are the same; what he establishes is that the morality of actions cannot be identical with, or be derived from, their rationality (458).

That Hume confused the two positions is probably because he thought that reason could discover an action to be immoral if and only if its immorality could be derived from, or consisted in, its being irrational. This is not an unnatural mistake to make; after all, it must be reason that discovers what is irrational, so if reason discovers immorality, immorality must, or so it seems, be irrationality. But though, put like this, the argument seems not implausible, it is in fact quite absurd. One might just as well argue that, since reason discovers the composition of the hydrogen atom, then hydrogen atoms must be capable of being rational or irrational. One might just as well argue that, since reason discovers what is true or false, reason cannot discover whether the earth is round or flat, or being round or flat would consist in being true or false. Hume's argument, if pursued to its logical conclusion, would allow reason to discover nothing at all.

The mistake, of course, is that two things—discovering rationality and truth, and discovering the properties of the hydrogen atom or the morality of actions—have been presented as alternatives when they are in fact compatible, indeed, inseparable. Reason discovering the properties of the hydrogen atom is not reason finding out things about the rationality or truth of the hydrogen atom, but its finding out things about the rationality and truth of the *propositions* describing the hydrogen atom. It is these which are true or false, and which it is reasonable or unreasonable to believe; discovering which propositions about the hydrogen atom are true and which false, reasonable and unreasonable, is one and the same thing as discovering things about the hydrogen atom, and does not in the least imply that the hydrogen atom must be true or false, reasonable or unreasonable. The case, for all Hume shows to the contrary, could be just the same with morality. For reason to discover what actions are immoral does not at all mean that being immoral is being irrational. All it means is that the *propositions* attributing immorality to the actions must be capable of being true or false, rational or irrational. Hume's argument does nothing whatsoever to show that they cannot be either or both of these things. His mistake is so bad that, where this argument is concerned, one almost feels inclined to accept his own low assessment of the *Treatise*, though I think this would be a mistake (he did not use this second argument in the more mature *Enquiry concerning the Principles of Morals*).

III

INTERLUDE: HUME'S GENERAL EPISTEMOLOGY

BEFORE we consider Hume's third argument to show that distinctions are not derived from reason, we must consider Hume's epistemology in general. Such a treatment, which must inevitably be brief, is best conducted by observing an old-fashioned distinction between conception, judgement, and argument—a distinction, incidentally, of which Hume himself disapproved (96 n.).

Hume's account of conception (Book I, Part I)

Hume's account of conception is as follows. As we have seen, he thought that the contents, operations, or 'perceptions' of the human mind fell into two groups, impressions and ideas (1). The difference between the two was that (a) ideas are fainter, or less forceful and vivacious than impressions (1–2). ('Less forceful and vivacious' are Hume's words for the difference between my visual mental image of a horse and my actual perception of a horse, and they seem to be as good as any.) (b) Ideas are copies of our impressions, for example, my mental image of a horse is a copy of my perception of a horse (2). More accurately, ideas are divided into simple and complex ideas (3). My idea of a golden pavement is a complex idea, consisting of simple ideas (of golden and of pavement?), and though it itself is not a copy of any impression, the simple ideas of which it is composed are. (c) My ideas are caused by and derived from my impressions, rather than the other way round (4–5); we must have an impression of redness before we can frame the idea, and if we have never had an impression of the taste of pineapple, neither can we have the idea of it. Hence Hume's principle *'that all our simple ideas in their first appearance are deriv'd from simple impressions, which are correspondent to them, and which they exactly represent'* (4).

Hume classifies impressions into those of sensation and those of reflection (7–8). Impressions of sensation—colour, shape, sound, heat or cold, thirst or hunger, pleasure or pain—first strike the soul 'from unknown causes', and then, when the ideas which are copies of these impressions return upon the soul, they produce impressions of reflection, such as desire and aversion, hope and fear (8–10). These are called impressions of reflection, because they are derived from ideas (though they are not copies of ideas, but original impressions).

Hume's distinction between impressions of sensation and impressions of reflection is obscure. Impressions of colour, shape, size, taste, sound, pressure, and feel (e.g. roughness) are surprisingly omitted, though they cannot but be impressions of sensation. Thirst and hunger, pleasure and pain, can be produced by our ideas almost as easily as can impressions of desire or aversion. And Hume is, quite likely, confusing the empirical thesis that hope is caused by our ideas with the conceptual thesis that hope involves an idea, for what we hope is that such-and-such is, was, or shall be the case.

Hume thought that all ideas, like all impressions, were completely specific and determinate (18). It is impossible either for us to frame an idea, or to come across an impression, of something which is coloured, but no determinate shade of colour; which is triangular, but neither isosceles, scalene, nor equilateral; which is numbered, but which is not an instance of any particular number. From this he concluded (as did Berkeley in the Introduction to the *Principles of Human Knowledge*) that abstract ideas, according to an account like Locke's, were impossible (20). Locke (in *An Essay Concerning Human Understanding*, Book III, Chapter III) regarded an abstract idea (say, of a horse) as an idea which possessed only the characteristics which horses have in common, and none of the characteristics whereby they differ, and so our abstract idea would have to possess determinable characteristics, without possessing these in any determinate form. For example, our abstract idea of a horse would have to possess size (for all horses are some size or other) but no determinate size (for not all horses are the same size) and colour (for all horses are some colour or other) but not any determinate colour (for horses differ in colour). This, Hume thought, was impossible.

Hume's account of our understanding of the meaning of a general word like 'horse' is then as follows (20–4). Such a word cannot have meaning because it calls to mind an abstract idea or image of a horse, for there are no such abstract ideas or images. It has meaning because it calls to mind a particular idea of a horse, say, a piebald stallion of fifteen hands, which serves as the representative of all the other possible particular ideas of horses. Similarly, a parliamentary constituency, members of which cannot all crowd into the House of Commons, may send a representative there. When we frame such an idea, we are conscious that it is just one of a range of particular ideas, each one of which might have served equally well, and so, when reasoning about it, we disregard those features which other particular ideas belonging to the same range would not share. Thus, should we hear the word 'triangle', and use the idea of, say, an equilateral triangle to represent the whole range of ideas of triangles which this word might equally well have called to mind, we are prevented from concluding that the three angles of a triangle are equal to one another because, should we do this, other ideas of triangles which are not equilateral are ready to crowd into the mind to correct this erroneous judgement (21–2).

Comments

(1) I have represented Hume as holding that, when we hear the word 'triangle', and it calls to mind a mental image or idea of a triangle, this idea represents all the other ideas of triangles which we might have had in the sense in which a Member of Parliament represents his constituency. It may have been his view, or, at any rate, he did not clearly distinguish this view from the view, that the idea we do have represents the others in the sense that it stands for them, or is a sign of them. It is, however, the word 'triangle' that is a sign in this sense, and it is not a sign of ideas of triangles, but of triangles. To the extent that we use mental images or ideas as words, in the way we frequently do when we are thinking to ourselves as opposed to reading or writing or hearing people speak, it is not necessary for ideas to resemble that for which they are a sign, as it is not necessary for our word for a horse to resemble a horse.

(2) The function of the mental images we call to mind in response to seeing, hearing, or writing a word, as opposed to the image words or idea-words we use to think *with*, seems to be to serve in lieu of the object the word means. If we are talking or thinking of triangles, and there are no triangles about, we can produce mental images or ideas of triangles to act as substitutes for them. It might be argued that some of these ideas or images are just instances of what we are talking about, as will be the case if our image or idea of a square is square, or our image or idea of a red thing is red, but this will certainly not always be the case. Our image of a horse is not itself a horse, nor is our image of a

fast-moving car fast-moving, or our image of a heavy weight heavy. We understand the meaning of a word when we are able correctly to apply it to the situations it describes, for example, to say 'horse' in the presence of horses and 'square' in the presence of squares. However, our understanding of words also manifests itself, in a subsidiary and derivative way, in our ability to frame mental images of horses when we heard the word 'horse', and of squares when we hear the word 'square'.

(3) Hume, unfortunately, pays very little attention to logical words like 'if ... then', 'either ... or', 'and', 'all', 'some', 'many', 'most', 'not', 'nothing', and so on. Clearly our understanding the word 'not' is in no way manifested by our being able to frame images of states of affairs which one can describe as not, and 'nothing' is not a name for a special kind of thing, which we might, albeit obscurely, image; it would be a mistake to try to frame an idea of the nothing which works faster than Anadin. Hume must also have trouble over 'abstract' words like 'justice'. How does one frame an image or idea of justice, as opposed to framing images or ideas of judges or lawyers or policemen, for example? The same is true of a vast collection of words and phrases like 'quadratic equation', 'parliamentary government', 'international law', 'literary style', or 'business acumen'.

(4) Hume could not have a solution to this problem since he was not aware of it. What he should have said is this. Although we cannot frame a mental image or idea of what is meant by the word 'not', we can frame mental images or ideas of the situations which would make true the sentences in which the word 'not' occurs. We can frame mental images or ideas, which serve as representatives of a whole range of images, which are images of what it would be like to come across a state of affairs which would make true, say, the proposition that there is no cheese in the larder, or that there is no justice for women, or no peace for the wicked. He could then have given an account of the meaning of these words as follows: to understand the meaning of a word is to understand whole sentences in which this word occurs, and to understand these sentences is to know how to recognize states of affairs which would make uttering them express something true, and, derivatively, to be able to frame images or ideas of these states of affairs.

(5) Hume, of course, if he is to escape the conclusion that words like 'virtue', 'vice', 'good', 'bad', 'right', 'wrong', 'duty', and 'ought' are meaningless, is committed to holding that there are ideas which these words call to mind, and impressions from which

these ideas are derived. It is perhaps not improper to anticipate that he may have some difficulty over doing this. One would not, for example, expect to have ideas, in the sense of mental images, of oughts, any more than one could have ideas of nots. One might argue that one's idea of a man's being good does not in any way differ from one's idea of the same man's being bad, and that one's idea of an action's being right does not differ from one's idea of the same action's being wrong. 'Good' and 'bad', 'right' and 'wrong' do not seem to show up in our images of a man in the way in which 'black' and 'white', for example, do. Hume himself, as we shall see, was well aware of this problem, and it was one reason why he thought that we must derive the ideas of virtue and vice from internal impressions of reflection, viz. feelings of approval or blame.

Hume's account of judgement (94–8)

Hume's account of judgement is to be found in Part III, Section VII of the *Treatise* (94–8). An account of belief must fall into two parts, an account of what is believed, and an account of what it is to believe it. What is believed is always that so-and-so, or that so-and-so is the case, except when what is believed is a person, which simply means that this person says so-and-so, and so-and-so is believed. We also speak of believing *in* something, where what we believe in is not that so-and-so, but cold baths or God, but it is unnecessary to go into the question of the proper analysis of belief in, and whether, like believing a person, it can be reduced to believing that so-and-so is the case. (I believe that it cannot be so reduced, at any rate, not without remainder.)

Hume held that the object of belief, or what is believed, is an idea (95). Since different people can believe, or disbelieve, the same idea, or just think about it without either believing it or disbelieving it, the difference between believing something and not believing cannot consist in a difference in the ideas of the believer and the disbeliever. In that case they would not be believing or disbelieving the same thing. It must consist in the *manner* in which these ideas are conceived (96). An idea which is both conceived and believed is, Hume thought, more forceful and vivacious than one which is conceived without being believed, or conceived and disbelieved; '... ideas, to which we assent, are more strong, firm and vivid, than the loose reveries of a castle-builder' (97).

Comments

(1) It seems to me that it is after introspection obvious that when we believe that a certain situation obtains, our ideas of it are not necessarily more forceful and vivacious than they are when we simply merely imagine what it would be like for this situation to obtain. It has been held, however, that by 'force and vivacity' Hume was not referring to an introspectable pictorial quality of our mental images; that he did not regard a belief simply as a vivid mental image. I shall say more about this later (p. 42).

(2) If Hume is right in thinking that all ideas are fully determinate or specific, he must have some difficulty in identifying them with propositions, which are always in some degree or other indeterminate or unspecific. Take, for example, the proposition that the cat is on the mat. All this asserts is that there is a cat of some shape or size or sex on a mat of some colour or material or origin. Any image or idea of a cat on a mat, however, must be fully specific, and so determine, improperly, the shape and size of the cat and the colour and pattern of the carpet.

(3) The difficulty Hume must have in giving an account of syncategorematic words like 'all', 'some', and 'not' must also reflect upon his identification of propositions with ideas or mental images. No particular idea can be identical with a proposition, which must be to some extent unspecific. The best Hume could do would be to identify the proposition that the cat is on the mat, not with one idea or mental image of a cat with a determinate shape and size on a mat of a determinate pattern, but with a range of ideas of cats on mats, having nothing in common but the fact that they are ideas of cats on mats, and differing in every other particular. No particular idea can be identical with the proposition that all men are mortal; it is not good enough to have an idea of one determinate man dying on one specific occasion. Even if we frame images of many men dying, what makes us say that this is the proposition that all men die, rather than the proposition that one hundred men, or at least a hundred men, or many men, die? In order adequately to express the thought that all men die, we must have the thought, which it implies, that there are no men who do not die. But with what idea is this negative proposition identical? We can, of course, have the idea, or frame the image, of a man who is not dying—which, of course, is not to say he never will—but this idea or image must be accompanied by the thought that this is a representation of a state of affairs the possibility of

which we are excluding. This is just the kind of thought that cannot be identified with any mental image or idea; ideas must, so to speak, all be affirmative.

(4) Hume cannot give a satisfactory account of disbelieving things. If, when we merely conceive the possibility that the cat is on the mat, our ideas are simply less forceful and vivacious than they are when I positively believe that the cat is on the mat, what happens when we disbelieve that the cat is on the mat? Are our ideas in that case still less forceful and vivacious? If Hume had had an adequate account of the meaning of 'not', he could have tried to resolve disbelieving that so-and-so is the case, into believing that so-and-so is not the case. In this case, disbelieving that the cat is on the mat could be held to consist in having very unlively ideas of a cat on a mat, but in having forceful and vivacious ideas of a cat in some other place than on a mat.

(5) This leads us to the fifth, and also absolutely conclusive, objection to Hume's theory of believing that so-and-so. Our ideas (as he himself pointed out when arguing against the existence of abstract ideas) are always fully specific, but our beliefs usually, if not always, are unspecific in some degree or other. We can, for example, believe that there is not any cheese in the larder without having any opinion at all about what, if anything, *is* in the larder. Our idea or image of a cheeseless larder must, however, be an image of a larder with nothing in it, or butter, milk, and eggs in it, or eggs, jam, and flour in it, and so on. Worse, our ideas, if they are ideas of a larder with butter in it, must be ideas of butter with a certain shape, and the larder itself be a larder of a certain kind, whereas we may have no wish to commit ourselves to anything about the nature of the larder and its contents over and above the fact that, whatever the larder is like, and whatever its contents are like, there is no cheese in it. It would be absurd to suggest that some of our ideas of the larder, those representing features of the larder to which we wish to commit ourselves to its having, are forceful and vivacious, and the others not. Since we might wish to commit ourselves to the existence of the larder, but not to its colour and shape, this would mean that we had to frame an idea of a larder which was forceful and vivacious, though the idea of its colour and the idea of its shape were not, and this is obviously impossible.

(6) It is an interesting feature of Hume's theory of the objects of belief that it precludes him from giving any account of contradictory beliefs. Berkeley (loc. cit.) advised his readers to think

with ideas, because this would make it impossible for them to talk (or think) nonsense. He was, of course, quite right to believe that one cannot think nonsense in ideas. Nonsensical strings of words cannot produce appropriate images. In suggesting that we confine ourselves to ideas, however, Berkeley was throwing out the baby with the bath water. For if we cannot entertain logical impossibilities, we cannot entertain logical necessities either. Unless we can think to ourselves that two and two might equal five, we cannot think that it is impossible that they should not, and so cannot think that they must equal four. It is impossible to distinguish, by ideas alone, between the thought that two and two do equal four, and the thought that they must. Hence, if we were confined to them, our theoretical knowledge would be greatly impoverished.

Hume's account of argument: (a) Demonstrative argument (69–73)

Hume's failure to distinguish between the object of a belief (a proposition) and a mental image must inevitably colour his account of demonstrative argument. Demonstrative argument would normally be explained as being based on a relation of entailment holding between those propositions which are the premises of the argument and those propositions which are its conclusion. This relation must be such that the premises of the argument cannot be true, and its conclusion false. Hume, of course, has to explain entailment as a relation between ideas.

There are, he thinks, seven kinds of 'philosophical' relation between ideas (14–15). These are resemblance, identity, relations of space and time, relations of quantity or number, degrees in quality (e.g. being brighter than), contrariety, and cause and effect. He thinks that only four of these give rise to demonstrative argument. These four are resemblance, proportions in quantity and number, degrees in quality, and contrariety (70).

The reason why these four relations give rise to the possibility of a demonstrative argument, whereas the other three do not, is this. If I frame an idea of four dots, and an idea of eight dots, I cannot conceive of the second idea as containing anything other than twice as many dots as the first idea. I can, on the other hand, conceive or frame an idea of four dots being either spatially adjacent to eight dots, or far removed from them. Hence I can know that four is the half of eight simply by attending to my ideas, but I cannot know how an object is spatially related to another object without having recourse to observation.

Comments

(1) It is not clear why Hume spoke of 'philosophical' relation, or whether he thought that these seven philosophical relations were the only kinds of relation that there were. If he did think this, he was wrong, for this list takes no account of relations such as being superior to (in the hierarchy of some organization), being the legal guardian of, being married to, entailing (a relation which holds between propositions), being instantiated by (a relation which holds between a universal like whiteness and a particular white thing which is an instance of it), being married to and being the father of (though perhaps this might be regarded as a species of the relation of cause and effect). The relations most conspicuous by their absence are relations such as being the legal guardian of or being the Member of Parliament for Smith's constituency. The reason why Hume omitted them, I suspect, is that they cannot easily be represented by our ideas. We may frame the idea of a man, but how is this idea altered when we frame the idea of his being a Member of Parliament, or a husband? Husbands or Members of Parliament do not necessarily look any different from any other men.

What is special about the seven philosophical relations? It may be, I suppose, that the seven philosophical relations are ones which might hold between ideas as well as between things; both ideas and things may be related by proportion in quantity and number, or by degrees of a quality. But then, Hume puts cause and effect down on his list of philosophical relations, and cause and effect is a relation which holds between objects or events rather than between ideas. We may have an idea of a spark's causing an explosion, but this does not mean that our idea of the spark causes our idea of the explosion, any more than our idea of a husband is married to our idea of his wife.

It could be that Hume supposed that, though there were more than these seven philosophical relations which could hold between objects, the seven philosophical relations were primary, in the sense that all other relations must be reduced to them. Any other relation between things will not be detectable (from which it follows that it may not even be meaningful to talk about such a relation) if it makes no difference to our impressions. It does not follow from this that the impressions which reveal the relation will have to be related by the same relation that they reveal. For example, being seven miles from is a relation (of space and time) between objects, and it is certainly revealed to us by our impressions or ideas of these objects. However, when we have impressions or ideas of an object which is seven miles from another object, neither our impressions nor ideas will be seven miles from one another.

(2) We have seen that Hume went astray when he treated our concept of a horse as being just *one* idea or image of a horse. Rather, he should have treated our having a concept of a horse as, over and above our having a disposition to recognize horses when we come across them, a further disposition to frame any one of a wide range of ideas, all of which would be ideas or images of horses. He went equally astray when he treated ideas or images as propositions. Our entertaining the proposition, or conceiving the possibility, that the cat is on the mat cannot consist in our having just one idea of a cat on a mat. It must consist in our having, or being disposed to have, any one of a wide range of ideas of cats of all shapes and sizes and breeds on mats of all shapes and sizes and makes. And anyone who can say to himself, *sotto voce*, the words, or some convenient private mental shorthand for them, 'The cat is on the mat' can be said to be entertaining this proposition, provided he can tell when a cat is on a mat, whether he is capable of having mental images of cats on mats or not.

(3) This emendation to Hume's account of conceiving a proposition must bring about a consequential amendment to his account of the seven philosophical relations, and in particular to those four which give rise to the possibility of demonstrative argument. Since there is not just one image which we must frame when we conceive of four cats, and not just one image that we must frame when we conceive of eight cats, we cannot speak of the relation 'twice as many' holding between *the* images which we have when we conceive of four cats and eight cats. Rather we should speak of this relation as necessarily holding between *any* of the images which we would frame when thinking of four cats and eight cats. If it were possible to frame any mental image of four cats which was not an image of half as many cats as a mental image of eight cats, it would be possible for four cats not to be half as many cats as eight cats. Similarly we must not suppose that cats are necessarily the same colour as dogs because we frame an image of a cat which is the same colour as the idea we frame of a dog. By an imaginative experiment we can replace the mental image of a black cat with one of a white one, or the idea of a black dog with one of a brown one. Hume's statement that two ideas are necessarily related by a given relation, if you cannot change the relation without chang-

ing the idea, might then be interpreted to mean that propositions are necessarily connected, by the relation of entailment, if any of the images or ideas of what it would be like for the first proposition to be true would *ipso facto* be images or ideas of what it would be like for the second proposition to be true. Obviously this fact is dependent upon the more fundamental fact that, if one proposition entails another, this must be because any situation which would make the first proposition true would *ipso facto* make the second proposition true.

(4) If we can define entailment in the way I have suggested, this avoids a serious difficulty with Hume's account of demonstrative inference as it stands. This difficulty is that, since he confuses propositions with ideas or mental images, he confuses entailment, which is a relation between propositions, with relations such as resemblance between mental images. Entailment cannot be the same as any relation which holds between ideas or mental images for two reasons. Firstly, those relations between images which Hume thinks give rise to entailment are different on different occasions. Sometimes it is one of resemblance, at other times it is proportion in quantity and number, and so on, but entailment is always the same. Secondly, entailment must be defined in terms of truth. One proposition entails another if the truth of the first necessitates the truth of the second, but it does not make sense to speak of ideas or mental images as being true or false, or not when, as in this context they should, these words mean 'is the case' and 'is not the case'. This is because ideas are as much 'real existences', in Hume's language, as anything else, and though they can be copies of impressions (and so, perhaps, copies of the object, if there is one, of which these impressions are impressions) 'being a faithful copy of' is not the same thing as being true. A map, for example, can be a faithful copy of its original, but it does not make sense to speak of a map as being true or false, asserted or doubted or denied, nor can a map figure as the antecedent or the consequent of a hypothetical proposition, or as a disjunct, or as a conjunct. The same is true of photographs or portraits or paintings. What can be asserted, doubted, or denied, etc. must always be that so-and-so, which mental images or ideas, maps, photographs, or portraits are not. Similarly, that something or other is square entails that it has four sides, but that it is square does not resemble that it has four sides, nor is that it is square related to that it has four sides by any other of the four relations which Hume thinks give rise to demonstrative argument.

(5) The fact that Hume confused entailment with a relation between ideas or images means that, in saying that only the four philosophical relations he mentions can give rise to demonstrative inference, he was making a mistake. That it *is* a mistake so to confine deductive inferences to these four relations is shown by the fact that, from '*A* is to the left of *B*', and '*B* is to the left of *C*', it is possible to deduce '*A* is to the left of *C*'. The relation of being to the left of it is a relation of time and place, and is not one of the four. Hume thought that such inferences could not be demonstrative because he thought (rightly) that we could conceive the idea of *A* as standing in any relation to the ideas of *B* and *C*. He forgot, however, that if we did frame an idea of *A* to the left of *B*, and then framed an idea of *B* to the left of *C* (or, better, framed an idea of *A*-to-the-left-of-*B*-to-the-left-of-*C*), then this would have to be an idea of *A* to the left of *C*.

What this comes to is this. Though Hume supposed that his account of ideas meant that only four of the seven philosophical relations on his list gave rise to demonstrative inference, there was no reason why this account should not permit of demonstrative inferences involving any of the seven philosophical relations he mentions. For example, if *A* is an early state of one and the same thing of which *B* is a later state, and *A* does not resemble *B*, it follows that *X* has changed; and if *A* is identical with *B*, and *B* with *C*, it follows that *A* is identical with *C*. Similarly, if anything is an effect, it follows that it has a cause, and if gold by definition dissolves in *aqua regia*, it follows that what does not dissolve in *aqua regia* is not gold.

The trouble is that if we simply appeal to our image or idea of an event, this will be the same whether this event is an effect or not, and, if we appeal just to our image or idea of gold, this will be unaffected by whether or not gold dissolves in *aqua regia*; gold will look just the same, whether it dissolves in *aqua regia* or not, and feel, taste, smell, and sound just the same, too. On the other hand, if we consider the *definition* of gold, or the *definition* of effect, we have no difficulty in seeing that gold must dissolve in *aqua regia*, or that effects must have causes.

This is not incompatible with the spirit behind what Hume says about demonstrative inference. If gold by definition dissolves in *aqua regia*, we will not be able to frame images or ideas of gold not dissolving in *aqua regia*. Such ideas will not be ideas of *gold* not dissolving in *aqua regia*, if the fact that a substance does not dissolve in *aqua regia* means that it is not gold. We think that we have formed an idea of gold when we have framed an idea of what gold looks like; but if gold (or anything else) possesses dispositional characteristics (like being soluble) or relational characteristics, such as being the object of avarice, these will not affect what gold looks like. Hence, to frame an adequate idea of gold, we must frame ideas of it standing in certain relations to other things, or behaving in certain ways.

(6) Sometimes, when considering whether any proposition is necessarily true, it is most natural to proceed in the way that is suggested by what Hume says about demonstrative inference, and try to see if you can frame ideas of what it would be like for it not to be true; and concluding, if you fail, that this is impossible. At other times, it seems more natural to proceed by seeing whether you can detect a contradiction in the sentence purporting to describe what it would be like for a necessary proposition not to be true. Doubtless, some people find the former method more natural, whereas others (those who are poor imagers) prefer the latter. In considering whether or not aunts can be anything other than female, the method of trying to image what it would be like if they were male works very ill; it is more natural to consider just what is meant by the word 'aunt'. At other times, for example, when a word has not got an explicit definition, it is easier to proceed in the former way. In the absence of any clearly formulated rule for the use of the word 'sailor', by reference to which you can decide whether or not sailors can be turned into swine, one might proceed by trying to envisage (or image) this state of affairs, and concluding, if one can, that it is possible for sailors to be turned into swine.

It might seem as if the two methods ought to give the same results. This must be subject to two provisos. Firstly, one's inability to image something must be due to its being impossible, and not to some limitation in our powers of imaging. Secondly, whenever we cannot image something, we must assume that this is impossible because any state of affairs which we do image will be one which cannot be properly described in any English sentence. *Prima facie*, however, there are cases when these two different methods do not give the same results. When you are considering whether space might have more than three dimensions, it is very difficult to image what it would be like for space to have, say, four dimensions, but not at all easy to see what contradiction is involved in the supposition that it has more than three dimensions. It is perhaps this fact that has led some philosophers notably Kant, to postulate a class of synthetic *a priori* truths, the necessity of which does not arise from the fact that their contradictories are self-contradictory. For myself, however, I am inclined to think that, if a certain state of affairs is impossible to image, then it should be possible to find, if we try hard enough, some logical impossibility involved in supposing this state of affairs to obtain; on the other hand, we may, if we exercise our powers of imaging fully enough, image what it is like for any non-contradictory supposition to be fulfilled.

(b) Inductive or probable argument (mainly Book III, Part III, Section VIII)

Hume devotes a large part of Part III of Book I of the *Treatise* to discussing probable or inductive inference. (Probable inference is to some extent a misnomer, for, as Hume himself says, 'One wou'd appear ridiculous, who wou'd say, that 'tis only probable the sun will rise tomorrow, or that all men must dye; tho' 'tis plain we have no further assurance of these facts, than what experience affords us' (124).

Obviously, impressions by themselves can give us but little knowledge of the world. We have no impressions of the past, or of the future, or of more of the present contents of space than are under our observation at any one moment. Even then, there is much more to any observed object than can be revealed by any one impression of it. Hence our impressions, which are fragmentary and intermittent, must be filled out with or supplemented by ideas. Not any kind of idea will do, for in that case there would be no difference between Rome, which is fact, and Mount Olympus, which is fiction. They must be ideas having a high degree of force and vivacity, i.e. they must be beliefs. This force and vivacity is conferred, Hume *often* says, upon them by impressions like them having been constantly conjoined—or what, according to Hume, amounts to the same thing, causally connected—with impressions resembling a present impression (102–3). For example, because smoke has always been found to be constantly conjoined with fire in my past experience, an impression of smoke both produces and confers force and vivacity on my idea of fire, and causes me to believe that there is fire to cause the smoke. Because chimney-stacks have been constantly conjoined with houses, an impression of a chimney-pot confers force and vivacity on my idea of the house under it (or, better, causes me to have a forceful and vivacious idea of a house under it). In this way my knowledge of the present contents of space is arrived at. Since I have observed a constant connection between fire at one time and embers at a later time, impressions of embers produce a forceful and vivacious idea of an earlier fire. Similarly, impressions of fire may produce forceful and vivacious ideas of later embers. In this way I acquire my knowledge of earlier and later times. Knowledge obtained by means of the testimony of others is in principle no different, for in this case we simply argue from the present impressions of print on paper—say, an account of the Battle of Hastings—to their remote causes—someone's witnessing and reporting on the Battle of Hastings (146). A memory image will serve as well as an impression as the premiss of an inductive inference. For example, we may infer that there was a Battle of Hastings as well from our recollection of reading about it as from actually witnessing the marks of print on paper (83, 89).

Comments

(1) Hume was quite wrong to *define* belief as 'a lively idea related to or associated with a present impression' (96). Even if this will do for beliefs which are sufficiently rational to be founded on a present impression, it excludes the possibility of our having irrational beliefs which are not so founded. Hume himself, inconsistently, says that madmen 'from an extraordinary ferment of the blood and spirits' believe everything that enters their heads; i.e., all their ideas are forceful and vivacious, whether they are related to a present impression or not (123). In any case, general beliefs such as that where there is smoke, there is fire, are not founded on a present impression, but based upon a constant conjunction between past impressions.

(2) Hume states: 'The only connexion or relation of objects, which can lead us beyond the immediate impressions of our memory and senses, is that of cause and effect; and that because 'tis the only one, on which we can found a *just* [my italics] inference from one object to another' (89). At other times he says that all inference from a present impression must be based upon experience of a constant conjunction between impressions, so that when I have a present impression which resembles either of these, it leads me to conceive a forceful and lively idea of the other (102). It is clear that Hume thought not only that there was a constant conjunction between a cause and its effect, but that all constant conjunctions between things contiguous in time and place were causal connections (93). Whatever may be said of the former view,

the latter is certainly mistaken. Experience of a constant conjunction between lightning and thunder may lead me to expect thunder when I see lightning (i.e. my impression of lightning produces a lively idea of thunder), but lightning does not cause thunder. Both, rather, are effects of the same cause.

(3) Hume usually talks of 'inferring' matters of fact, but it would be absurd to suppose that all matters of fact are inferred. For example, we may infer the 'matter of fact' that there is fire, but we infer this from the premiss that there is smoke, which premiss is not inferred. We perceive, rather than infer, that there is smoke, but that there is smoke is nevertheless just as much a matter of fact as that there is fire. Hume himself says, though he often carelessly forgets it, that ' 'Tis impossible for us to carry on our inferences *in infinitum*; and the only thing, that can stop them, is an impression of the memory or senses, beyond which there is no room for doubt or enquiry' (83).

(4) We have already seen that images cannot be beliefs, nor can they be believed. What is believed is always *that* so-and-so. And it seems fairly obvious that the images or ideas associated with what we believe are not always more lively than those which are not. It may be that, in using the words 'forceful', 'solid', and 'firm', as well as the word 'lively', Hume was trying to describe the mental attitude we take up to ideas (propositions), which he says we believe, rather than to describe the ideas themselves. (See H. H. Price, *Belief*, 1969, Lecture 7.) Certainly he does speak of the difference between belief and 'incredulity' as at least partly depending upon the effect the former has on our reasonings. We use beliefs as premisses, but otherwise argue only hypothetically, by pointing out what would follow, if something were true. We also act upon beliefs, whereas we do not act upon propositions that we do not believe. Nevertheless, he does think that ideas, when they are beliefs, resemble impressions in force and vivacity. Ideas, however, cannot resemble impressions in our having confidence in them, or in our being prepared to act upon them, or in our using them as premisses in our reasoning, for neither ideas nor impressions can be the objects of confidence, or be acted upon, or used as premisses.

What Hume might have said with great plausibility would have been that when we entertain a proposition which is not about a present impression, and believe it, our attitude to this proposition resembles our attitude to propositions about present impressions, which propositions we cannot help believing, if they are true. Unfortunately, he cannot himself accept this modification to his theory, for he has great difficulty in explaining how we can have beliefs about our present impression at all. A belief, according to Hume, is always a forceful and vivacious idea, whereas impressions are not ideas at all, and indeed, if the impression is present, it is superfluous to have an idea of it.

(5) Hume sometimes says that it is constant conjunction between impressions which causes us to have a forceful and lively idea of either when only one of the impressions previously associated is present. At other times he says that contiguity, resemblance, education, and 'extraordinary ferment of the blood and spirits' (123) can produce forceful and vivacious ideas (Section X). For example, it is because of contiguity that my belief in the New Testament is strengthened when I visit Palestine. It is because of resemblance that I find it easy to believe that one billiard ball will impart its motion to another billiard ball, and because of lack of resemblance that I find it difficult to believe in an after life or in the causation of mental events by brain processes. 'Education', in a derogatory sense, is Hume's name for what he supposed to be the fact that the frequent recurrence of an idea heightens its force and vivacity, and so causes it to be believed. It is clear that Hume usually thinks that only constant conjunction gives rise to rational belief ('a just inference from one object to another'), but this does not remove the contradiction; nor does his saying that the relation of cause and effect is requisite to give force to these other relations. He himself is aware of the inconsistency, and says (107), 'But as we find by experience, that belief arises only from causation, and that we can draw no inference from one object to another, except they be connected by this relation, we may conclude, that there is some error in that reasoning, which leads us into such difficulties.' I cannot find, however, in what Hume says subsequently that he does point out any error in this reasoning.

(6) As we have seen, Hume can give no satisfactory account of entertaining negative, hypothetical, disjunctive, universal propositions, or propositions involving 'some', 'most', 'few', or 'half the time'. The fact that Hume can give no satisfactory account of universal propositions is especially unfortunate, for one of his main aims in Part III of Book I of the *Treatise* is precisely to give an account of our knowledge of such propositions as 'Whenever there is smoke, there is fire' and 'Whenever one billiard ball hits another, it causes it to move'. Not only can Hume give no satisfactory account of entertaining universal propositions, his account

of belief will not apply to them either. For though my belief that there is fire may be linked to my present impression of smoke, my belief that whenever there is smoke there is fire is not linked to any present impression. I believe it just as firmly, and just as rationally, when I see smoke or when I do not.

(7) It is, of course, Hume's view that inductive reasoning is not rational. This is because it is not valid by the canons of deductive logic. All inductive inference presupposes that the future will resemble the past, for example, that billiard balls will impart their motion to other billiard balls in the future as they have in the past. There is no contradiction in supposing, however, that the future does *not* resemble the past, and our only reason for believing that it does is itself inductive, i.e. we believe that the future will resemble the past because, in the past, it always has done.

It is clear, however, that Hume was not against our making inductive inferences. He must work with *some* criterion of appraising non-deductive ways of arriving at beliefs, for, in condemning education and, in most cases, resemblance and contiguity, as compared with constant conjunction, he is clearly presupposing a criterion of rational appraisal, by the application of which the latter stands but the former falls. He also thought that nature had given us no alternative but to base our beliefs on constant conjunction and also, though this itself rests on an inductive argument, that nature would continue to give us no such alternative in the future.

(8) As we have seen, Hume fails to distinguish between believing, on seeing smoke, that there is fire, and believing that whenever there is smoke there is fire. He also fails to distinguish between believing that whenever there is smoke there is fire, and that smoke is caused by fire. He attempts to reduce causation to uniform connection; to believe that A causes B, according to Hume, is simply to believe that A is contiguous to B, precedes B in time, and that *whenever* an event like A occurs, an event like B also occurs. This seems to leave out the element of necessity in causal connection. Hume attempts to give an account of this necessity by saying that it is derived from the felt compulsion to pass from an impression or idea of A to an idea of B, which is the result of events like A and events like B always having been found together in the past (Section XIV).

In Book II, Part III, Sections I and II, he half hints at a more satisfactory theory of the nature of that necessity which is involved in a causal connection. It lies in its unavoidability. A cause is linked to its effect by the fact that I cannot separate the two, and in the fact that, given the one, there is nothing I can do to prevent the other. It makes no difference whether the events connected are physical or mental (natural or moral). 'The same prisoner, when conducted to the scaffold, foresees his death as certainly from the constancy and fidelity of his guards as from the operation of the ax or wheel' (406).

(9) As I have said, Hume disapproved of the division of 'acts of the understanding' into conception, judgement, and reasoning (96 n.). Judgement and reasoning, he thought, resolved into conception, and the only important distinction was that between conceiving something without and conceiving something with believing it. Hume is in a state of some confusion here. Obviously he cannot resolve judging into conceiving, for judging involves believing, but conceiving does not. Equally obviously he cannot resolve reasoning into conceiving, for reasoning often involves a passage from one belief to another, but a belief cannot be reduced to a conception, and a passage from one belief to another cannot be reduced to a belief.

At the back of Hume's mind, presumably, though it is not the reason he gives, is the idea that having a conception of a horse, entertaining the proposition (without believing it) that there are horses, and entertaining the proposition (which could form the basis for an inference or piece of reasoning) that, if there are horses, it cannot be the case that there are no horses, can all manifest themselves in the form of having an image or idea of a horse. But though they can all manifest themselves in this way, there are ways in which the conception can manifest itself, but the judgement not; and ways in which the judgement can manifest itself, but the reasoning not. I can have a concept of a horse, without having entertained the possibility that they might fly, and entertained this possibility without having ever considered whether, if they could, they would still be kept in paddocks.

Hume would have managed better if he had regarded conception, entertaining a proposition, and considering a relation between propositions not as individual ideas, but as ranges of ideas. Then the range of ideas which manifests our conception of a horse can be different from those ranges of ideas which manifest different propositions about horses, and these again different from ranges of ideas which manifest our understanding of different arguments about horses. Hence the ranges may be different, though some of the items in these ranges are the same. Any idea which manifests our understanding of an argument about horses will manifest our grasp of propositions about horses, and any idea which manifests our grasp of propositions about horses will be a manifestation of our having a concept of a horse, but the converses of these two propositions are not true. However, the difficulty that Hume has in distinguishing between concepts, propositions, and arguments (I prefer these expressions to 'judgement' and 'reasoning', because these introduce the irrelevant notion of belief) is just another reason for thinking that his attempt to understand these in terms of images or ideas alone is only partially successful.

IV

MORALITY NEITHER (A) SUSCEPTIBLE OF DEMONSTRATION NOR (B) A MATTER OF FACT (463-9)

H u m e ' s third argument to show that moral distinctions are not derived from reason is as follows. 'If the thought and understanding were *alone* [my italics] capable of fixing the boundaries of right and wrong, the character of virtuous and vicious either must lie in some relations of objects [he normally says "ideas"], or must be a matter of fact, which is discovered by our reasoning. This consequence is evident. As the operations of human understanding divide themselves into two kinds, the comparing of ideas, and the inferring of matters of fact; were virtue discover'd by the understanding; it must be an object of one of these operations, nor is there any third operation of the understanding, which can discover it' (463).

Since he thinks our apprehension of moral distinctions falls into neither of these two categories, Hume concludes that it is not the work of reason. In other words, if reason discovers moral distinctions, we must come to know them in one of two ways. We may come to know them in the manner in which we know that two birds are equal in number to two elephants, by comparing our ideas or images of each, and seeing that they must be equal in number. Alternatively, we may come to know them in the manner in which we know that there is, in reality, something answering to our idea of fire, or something of which our image of fire is a copy, by inferring this from our perception of smoke.

Comments

(1) Hume denies only that the understanding is *alone* capable of ascertaining what actions are right or wrong. He does not deny that the understanding has a part to play in this operation. Its function is to discover the matters of fact upon which, he later maintains, our moral sense pronounces judgement.

(2) It would have been nice if Hume had consistently used the

word 'understanding' for the generic process of either comparing ideas or inferring matters of fact, and 'reason' for the specific process of comparing ideas, but he does not.

I shall divide this chapter into two parts. In the first I shall consider Hume's reasons for thinking that morality does not admit of demonstration, and that our knowledge of morality is not arrived at by comparing our ideas. In the second I shall consider Hume's reasons for thinking that the morality of an action does not consist in any matter of fact.

A. MORALITY NOT SUSCEPTIBLE OF DEMONSTRATION (463–8)

Hume's argument to show that morality does not admit of demonstration is very simple. Morality neither consists in one of the four philosophical relations which Hume thinks give rise to demonstration, nor in any possible new relation, which Hume has omitted from his list.

Morality not one of the four philosophical relations which Hume thinks give rise to demonstration (463–4)

Morality cannot consist in one of the four relations which Hume recognizes as giving rise to demonstration (i.e. resemblance, contrariety, degrees in quality, and proportions in quantity and number) because these are all as capable of belonging to inanimate objects, of which morality cannot be predicated, as they are of belonging to actions, passions, and volitions, of which morality *can* be predicated.

Comments

(1) Hume's proof that morality cannot consist in any one of the four philosophical relations is absolutely conclusive. If to say that something was right or wrong, good or bad, virtuous or vicious was to say that it was like, unlike, redder than, or equal in number to something else, then anything which possessed the latter properties would also have to possess the former (for they would be the same properties). It is obvious that, where each one of these relations is concerned, it is easy to find things which possess them, but which do not possess any moral properties at all.

(2) We have already seen that Hume was wrong in supposing that only the four philosophical relations he mentions do give rise to demonstration, and that he wrongly supposed this because he confused propositions with ideas or mental images.

(3) Not only does Hume's proof of the conclusion that morality is not susceptible of demonstration fail, because one of his premisses, that only four relations between ideas give rise to demonstration, is false; his conclusion is also false, and rather obviously false at that. That if all rich men are immoral, and Croesus is rich, then Croesus is immoral, is just as good a necessary truth as is that, if all whales are mammals, and Figaro is a whale, then Figaro is a mammal. That every voluntary action is either right or wrong is, if true, a necessary truth, and, if it is a necessary truth, it is just as good a necessary truth as that every natural number is either even or odd. That if Tom is morally better than Dick, and Dick morally better than Harry, then Tom is morally better than Harry, is just as good a necessary truth as that if Timothy is older than John, and John is older than Susan, then Timothy is older than Susan.

(4) The reason why Hume overlooked the fact that there were necessary truths about morality was, I think, this. I said earlier (pp. 39–40) that sometimes it was easier to see whether a given sentence expressed a necessary truth by considering how the words in this sentence were to be used, and at other times easier to see this by trying to frame an idea of what it would be like for it not to be true. The former method works best with necessary truths such as that bachelors are unmarried. The latter method works best with necessary truths such as that what is coloured must be extended. Though it is very easy to show that there can be necessary truths about morality if one considers the definitions of certain words—that murder is wrong is a necessary truth about morality, for example, if 'murder' just means 'wrongful homicide', which it does—it is very difficult to show that there are necessary truths about morality by trying to envisage situations which would be described by, for example, permissible homicide. One is reluctant to conclude anything from the fact that one cannot frame an image or idea of morally permissible murder, because one has difficulty in framing an image of murder to start with. It is quite easy to frame an image or idea of one man killing another, but most people would use the same image or idea to represent someone killing someone else in self-defence, in battle, with provocation, without provocation, and so on; hence, how do you represent on your image, or frame an idea of, its being a murder, and so how do you, by considering your ideas, decide whether murder must be wrong, or may be right? I suggested above that a sentence may be held to formulate a necessary proposition if it was

impossible for us to frame an idea of what it would be like for it not to be true; but in this (as in many other cases) it is so difficult to frame an idea of its being murder at all (or of its being wrong at all) that this method is difficult to apply. (It would also be difficult to apply when trying to decide whether allegedly necessary propositions of non-Euclidean geometry are really necessary, for it is difficult for us to frame images—or construct diagrams—of what it would be like for them to be true.) If, however, you consider the second method of deciding whether a sentence formulates a necessarily true proposition, the method of considering whether the definitions of the words in these sentences make it contradictory to deny the propositions these sentences express, there is no more difficulty in seeing that sentences such as 'Murder is wrong' can formulate necessary truths than there is in seeing that sentences such as 'Bachelors are men' express necessary truths. It can be argued just as well that murder must be wrong, because murder is by definition wrong, as that bachelors must be male, because bachelors are by definition male.

(5) There is therefore no reason why there should not be necessary truths of morality, and, indeed, there are such truths. That Hume was wrong in thinking that there were no such things, however, does not make as much difference as might be supposed. Though there are necessary truths about morality, they all suffer from a defect, which makes them useless as a guide to action. The necessary truth, that if murder is wrong, and contraception is murder, then contraception is wrong, will not tell us whether contraception is wrong; though that contraception is wrong follows from the premisses, it is possible to dispute both the premisses that murder is wrong, and that contraception is murder. This is because necessary truths are all hypothetical, and simply tell you that, if you hold some things, you must also hold others; or, what comes to the same thing (since to say that one proposition necessitates another is the same as to say that the first proposition cannot be combined with the negation of the second), they tell you that if you hold some things, you may not also hold others.

It may seem as if you can get round this limitation of demonstrative reasoning in ethics if you not only argue that certain conclusions necessarily follow from certain premisses, but also argue that some premisses, from which conclusions necessarily follow, are necessarily true. For example, you may argue that not only is it necessarily true that if murder is wrong, men ought not to commit it; it is also necessarily true that murder is wrong. The statement that murder is necessarily wrong, however, only tells us that if any action is an instance of murder, it is wrong; it cannot be a necessary truth that any action is a case of murder. And the trouble with this necessary truth is that, if it is true, it is true only because murder is by definition wrong, which means that we will, or, at any rate, should, refuse to describe any action as murder unless we think that it is wrong. In this case, we cannot establish that an action is a case of murder, and is in consequence wrong, without begging the question. Unless we have *first* established that it is wrong, we are not entitled to describe it as a case of murder at all. Similarly, in the example given above, we cannot, if murder is necessarily and by definition wrong, establish that contraception is murder until we have first established that contraception is wrong, and so the argument: 'Murder is wrong; contraception is murder; hence, contraception is wrong', though valid, begs the question.

(6) It must not be supposed that this limitation upon what can be done by means of demonstrative reasoning in morality is a limitation which applies to moral argument only. Though the argument 'Smith is a bachelor, therefore he is unmarried' is valid, it begs the question, in that, if there are any doubts about whether or not Smith is married, there must be equally grave doubts about whether or not he is a bachelor. And though the statement that two and two are four is demonstrably true, it is, nevertheless, hypothetical; it can tell you that, if there are two men in one room, and two men in another, then there are four men in the two rooms, but it cannot tell you how many men there are in these two rooms in fact. The whole function of demonstrative reason, therefore, is to tell you what are the consequences of certain assumptions, and this is just as much a fact about demonstrative reasoning in ethics as it is about demonstrative reasoning in any other sphere. It would be ridiculous to conclude from this, of course, that demonstrative reasoning was unimportant. It is, on the contrary, enormously important to know what are the consequences of any assumptions which you happen to make, and these consequences may be far-ranging, surprising, and difficult to elicit. However, it is also most important to realize that, as Hume himself pointed out, demonstrative reasoning can never establish any matter of fact, and, though he was wrong in supposing that this fact imposed a more severe limitation on moral reasoning than upon reasoning of any other kind, we ought to be grateful to him for drawing our attention to it.

Morality not a new relation, other than the four philosophical relations Hume has mentioned (464–6)

Hume now considers the possibility that his list of the philosophical relations which give rise to the possibility of demonstration is incomplete. He argues that he cannot reasonably be expected to refute this view until it is explained what this new relation is, and challenges his opponents to produce a fifth relation, which both gives rise to demonstration and is such that morality can plausibly be held to be identical with it. However, he stipulates that this relation must fulfil two conditions, which, he thinks, is impossible.

The *first* of these conditions is that the relation his opponents produce must be capable of holding only between internal actions on the one hand and external objects on the other; it must not be possible for this relation to hold between one internal action and another, or between one external object and another (464–5). If it could hold between one external object and another, inanimate objects could be right or wrong. If it could hold between one internal action and another, actions would be right or wrong, irrespective of the circumstances in which they were performed.

The *second* condition is that the new relation, whatever it may be, must *necessarily* have the effect of moving any creature rational enough to apprehend it (Hume actually says 'every rational creature') (465–6). It is impossible, he thinks, that the apprehension of the existence of any such relation should have this effect. For one thing, Hume believes that he has proved that reason cannot alone move us to action, from which it follows that merely knowing that an action possesses a relation cannot have any effect on our actions at all. For another thing, whatever effect knowing or believing that an action possesses such a relation does have (say, in arousing a passion to perform the action in question) is something which can be discovered by experience only. This is because all beings in the world (and so, as in this case, the relation and the passion aroused by the relation) are 'entirely loose and independent' (466) (Hume means logically independent) of one another. This entails that from the existence of one thing (a relation) nothing at all can be inferred about the existence of another thing (a passion in the mind of someone who knows that an action has this relation).

Comments

(1) It is not immediately obvious that there are no relations which can hold only between passions, volitions, and actions on the one hand and other things. Sir David Ross, in *The Right and the Good* (1930), indeed, once held that rightness was itself such a relation, a relation of suitability, which held between an action and the situation in which the action was performed.

(2) It must not be supposed that because morality is not *identical with* any relation that holds only between 'internal actions' and 'external objects', that it follows that it *is not* such a relation. To do this would be like arguing that because yellow is not identical with red or green or black, and so on, there is no colour with which it is identical, and so it is not a colour. This mistake has some importance in the history of moral philosophy, for it has often been supposed that, because rightness and goodness are not *identical with* any natural characteristic, they cannot themselves *be* natural characteristics. This does not follow (even if it is true).

There is not much of a case, however, for saying that morality *is* a relation. (The statement that *morality* is a relation, is not just one statement, but the conjunction of the statement that rightness is a relation, goodness a relation, being virtuous a relation, and so on, and, of course, some of these statements could be true while others are false.) There is nothing which follows the words 'right', 'good', and 'virtuous', in sentences in which they occur, which is even remotely analogous to the words which follow after 'on', in sentences such as 'The book is on the table'. 'Good *for*' and 'good *to*', as opposed simply to being good, may be relations, but these are not senses of 'good' that Hume is concerned with. 'Right for Smith' is sometimes a very bad way of saying 'supposed to be right by Smith'. At other times it is a way of saying that some action would be right if performed by Smith when (because circumstances are different) it would not be right if performed by Jones. This no more involves a relational sense of 'right' than 'would not be green, if painted by Jones' involves a relational sense of 'green'.

(3) Why does Hume think that the *second* condition, that the new relation must necessarily move the will, has to be fulfilled at all? After all, one might say, rationalism is a view about how moral distinctions are apprehended; it holds that, by reasoning alone, we can work out, *a priori*, what things are right and wrong, just as we can, by reasoning alone, work out, *a priori*, what

theorems are true in mathematics or logic. It should be no part of such a view that such *a priori* reasoning necessarily has any effect at all, let alone that it must move us in one way, rather than in another.

Hume, it must not be forgotten, is committed to the view that the apprehension of a moral distinction moves us to action in a way in which apprehension of a matter of fact does not, i.e. the former alone moves us to action, without the co-operation of a passion, whereas the latter moves us to action only with the co-operation of a passion. From this it would follow that awareness of a moral distinction cannot consist in the apprehension that a relation is present, for such awareness, obviously, would only move us to action *with* the co-operation of a passion. However, there is no reason why Hume's rationalist opponents should not simply concede this, and maintain that there *is* a passion for morality.

In any case, Hume does not make use of the premiss, that the apprehension of a moral distinction alone moves us to action, in this argument. The premiss he does make use of is that it can only be a contingent fact that the apprehension of the presence of any relation moves us in the way it does, or, indeed, moves us at all. This, together with the additional premiss that if morality consists in relations apprehended by reason, morality must *necessarily* move us, would entail that reason does not apprehend moral distinctions. Is there, however, any reason why this second premiss should be true?

(4) One reason why Hume thinks that the apprehension of the presence of a relation cannot *necessarily* move us to action is that "'Tis one thing to know virtue, and another to conform the will to it' (465). This is true, despite what some contemporary moral philosophers have said about believing or accepting a moral principle consisting in acting on it. It *would* follow that reason does not apprehend moral distinctions, if it *were* the case (which it is not) that if one believes that reason apprehends moral distinctions, one must also believe that apprehending morality necessarily has an effect on the will. What Hume does not seem to see is that this very same contention, that to know what is right is not the same thing as to do it, is at least *prima facie* inconsistent with his own view that awareness of a moral distinction alone moves us to action.

(5) In this paragraph Hume suggests that it is, or has been held to be, a corollary of the view that moral distinctions are apprehended by demonstrative reasoning that "'tis not only suppos'd,

that these relations, being eternal and immutable, are the same, when consider'd by every rational creature, but their *effects* are also suppos'd to be necessarily the same; and 'tis concluded they have no less, or rather, a greater, influence in directing the will of the deity, than in governing the rational and virtuous of our own species' (465). Here there is a suggestion that if morality, because apprehended by demonstrative reasoning, consists of a relation, it must necessarily influence the will of the deity, and possibly also a suggestion that, if morality is not apprehended by a process of demonstrative reasoning, then it will have no necessary influence on the deity. Both these suggestions are false. It is, of course, true that, *if* morality necessarily influences the will of any being who can apprehend it, it necessarily influences the will of the deity; but Hume has not shown that, if morality consists in relations, it does necessarily influence the will of anyone capable of apprehending these relations.

(6) I do not know whether or not Hume thought that morality did necessarily have an influence on the will of the deity. Probably he did not. However, if anyone were to suppose that morality does have a necessary influence on the will of the deity, they would be guilty of a not uncommon logical mistake. It is true, of course, that, because God is, by definition, perfectly good, his will necessarily is guided by the moral law, but this just means that it is a necessary truth that it is guided, not that it is in fact necessarily guided, nor that it is a necessary truth that it is necessarily guided. (The proposition that is true, if we substitute p for 'x is God', q for 'x's will is governed by the moral law', C for 'implies', and L for 'is necessary', is the proposition symbolized by $LCpq$; the propositions which are false are $CpLq$ and $LCpLq$. $LCpq$ is a *necessitas consequentiae*; $CpLq$ is a *necessitas consequentis*; $LCpLq$ is both a *necessitas consequentiae* and a *necessitas consequentis*.)

(7) What, Hume says, must be the case if the awareness of moral distinctions is to consist in an apprehension of a necessary truth is that the relations in which morality must then consist must guide the will of 'the *virtuous* of our own *species*' or influence 'every *well-disposed* mind' (465). (In both quotations the italics are mine.) Inserting the words 'virtuous' and 'well-disposed' has the disadvantage of making the proposition Hume wants to prove nearly tautological, for it could be argued that, to the extent that one is not guided by morality, one is *not* a virtuous creature, and one's mind is *not* well-disposed. The proposition which Hume should have tried to prove is the proposition that, if morality is apprehended

by demonstrative reasoning, the relations in which it must then consist must necessarily have an influence on any mind capable of apprehending them.

(8) In this paragraph occurs the remark 'In order, therefore, to prove, that the measures of right and wrong are eternal laws, *obligatory* on every rational mind [note that here Hume does not say virtuous or well-disposed mind], 'tis not sufficient to shew the relations upon which they are founded: We must also point out the connexion betwixt the relation and the will; and must prove that this connexion is so necessary, that in every well-disposed mind, it must take place and have its influence; tho' the difference betwixt these minds be in other respects immense and infinite' (465). Is Hume right in thinking that, in order to show that something is everywhere obligatory, one must also show that it must always have an effect on the minds of rational creatures? (He does not here make it clear whether he simply means some effect, or enough effect to make them do what is obligatory, though the latter is most implausible.) It could be that he thought that it is only if morality consists in a relation that it must have this effect, if it is to be obligatory. It is, however, more likely that he thought that, since morality, whether it consists in a relation or not, must have an effect on the mind if it is to be obligatory, it cannot be a relation, for, if it were, it could not have such an effect.

But is it true that what has no effect on the mind cannot be obligatory? The assertion that what has no such effect cannot be obligatory is ambiguous. It may mean that I cannot assert that some person—who may be myself—is under an obligation to do something, unless what he has an obligation to do has an effect on *his* mind, or it may mean that I cannot assert that anyone has an obligation to do something unless what I say he is obliged to do has an effect on *my* mind. Quite often Hume is thinking of the former contention, as I believe he is in this passage. But quite often it is the latter contention that is at issue, for if a sentiment is necessary, if we are to make moral judgements, it is a sentiment whose function it is to make the *judger* not indifferent to the action being said to be obligatory; it is not then its function to make the agent not indifferent to the action which is judged to be obligatory. For example, when Hume says (E290–1) that we condemn Nero for performing actions which arouse detestation in us, although they arouse no similar feelings in him, it must be this second view that his remarks presuppose; Nero was obliged to refrain from the things that Nero did, although they did not repel him, the

agent. In the special case when the agent is making moral judgements about his own actions, it follows that he cannot make the
moral judgement that an action which he is contemplating omitting, is obligatory, unless he feels attracted to it (or repelled from
omitting it). Hence, Nero could not judge that his own actions
were wrong, and hence, if it is a necessary condition of our judging
that Nero's actions were wrong that Nero himself should judge
them to be wrong, then we, too, cannot judge his actions to be
wrong. Hume, of course, thinks that we can judge Nero's actions
to be wrong, which is apparently an inconsistency.

There is little to be said, as we shall see (pp. 112–13), for the view
that the reason why a man cannot judge an action to be right unless
he is attracted to it, or wrong unless he is repelled by it, is that
such judgements are judgements *about* his own feelings of attrac-
tion or repulsion. There is more to be said for the view that it
is improper to say that an action is obligatory unless one feels some
repulsion (disapproval) towards its omission. The difference
between these two views is as follows. If to judge that something
is obligatory is to judge that it attracts me (or that its omission
repels me) in some way, and I am not attracted (or repelled), then
my judgement is false. If, however, I ought not to say of an action
the omission of which does not repel me that it is wrong, then
my judgement is not false, but is misleading in some other way.
For example, I ought not to say 'It is raining' if I do not believe
that it is raining, and anyone hearing me say 'It is raining' is
entitled to infer that I believe that it is. If I say that it is raining
when I do not believe that it is, my judgement may, nevertheless,
be true, for it is, obviously, possible for it to rain, even though
I do not believe that it is raining.

Hume's examples (466–8)

At this point Hume attempts to reinforce the argument he has
just given by means of two examples. He thinks that, if morality
consisted in relations, it would follow that certain things which
happen to vegetables and to animals would be immoral, although
in fact they are not. (If morality does not consist in relations, *a
fortiori*, it cannot consist in any of the four philosophical relations
which give rise to demonstrative argument.) If morality consists
in relations, then the suffocation of an oak by the tree which has
grown up from one of its own acorns would be as immoral as when
a human being kills its own parent, for the same relations are
present in both cases. For the same reason, it would be as immoral

for an animal to copulate with its parents as for a human being
to commit incest.

To meet two possible objections, Hume argues that, in the first
example, it can make no difference that a human being *deliberately*
kills its parents, while a young oak does not deliberately kill its
parent tree (467). If the immorality consists in relations, and these
are the same in both cases, it will not matter whether they are
produced deliberately or not, so long as they are there. In the
second example, it does not matter that the human being has the
power to discover the immorality of copulating with one's parent,
whereas an animal has not this power. If morality consists only
in relations, and there is no difference in the relations which obtain
in the two cases, there can be no relevant difference between these
two cases for the reason of the humans to discover (468).

The point that, if morality consists in relations, it makes no difference whether they are produced deliberately or not, is made
when Hume discusses his first example, parricide. The point that,
if morality consists in relations, and there is no difference between
the relations possessed by two things, one cannot have a knowledge
of the morality of the one thing but not of the other, is made when
discussing the second example. There is no reason, however, why
both points should not be made about both examples.

Comments

(1) Hume sometimes seems to be arguing that there is no difference between the relations in which animals stand to their
parents, and between the relations in which human beings stand
to their parents. This is puzzling. One would have supposed that
what he ought to have been trying to prove was that there is no
difference between the relations possessed by the *action* of a tree
killing its parent and the *action* of a human killing its parent; no
difference between the relations possessed by the *action* of an animal copulating with its parent and the *action* of a human copulating with its parent.

(2) If you limit yourself to Hume's seven philosophical relations
(resemblance, identity, relations of time and place, proportion in
quantity or number, degrees in any quality, contrariety, and causation) it *is* difficult to think of any of these relations, which are possessed by the human actions, but not by the vegetable ones. There
are, of course, many other differences between the animal and
vegetable 'actions' on the one hand, and the human actions on
the other. The human actions are performed by someone who

knows roughly what he is doing, and by someone who both needs and is capable of having a social organization which is dependent on the existence and enforcement of rules (some of which will almost inevitably prohibit parricide and probably prohibit incest). What Hume says suggests, though the suggestion may be unintentional, that if the difference between the human actions and the animal and vegetable ones consists in knowledge which the human possesses, but the animal or vegetable does not, this difference must consist of knowledge of those relations in which the immorality of their actions consist. Then he argues, correctly, that the *only* difference between the two cannot be that the human knows the immorality of what he is doing, while the animal does not. Knowledge of a difference presupposes some difference which this knowledge is knowledge of. But this does not mean that knowledge of *other* differences between two actions cannot make one wrong and the other right. For example, knowledge that the woman with whom you are copulating *is* your mother may make it immoral to copulate with her, even if it is true that knowledge that it is *immoral* to copulate with her cannot be what makes it immoral to copulate with her. Even this latter has been disputed.

(3) It may well be that the relation signified by the words 'mother' and 'father' is a causal one, and in this respect, in broad outline, at any rate, no different from the relation in which an oak stands to the tree which bore the acorn from which it sprang, and that in which an animal stands to its parent. There is, however, a social difference between humans, on the one hand, and animals and vegetables, on the other, which Hume here overlooks. In most places it is a man's biological parents who are allocated by society the task of caring for him until he is capable of caring for himself. There is no reason, however, why the two things necessarily go together, and in some societies they do not. For example, the social tasks that in England are allotted to one's biological father are, in Melanesia, allotted to one's biological uncle.

B. MORALITY NOT A MATTER OF FACT (468–9)

Hume has proved, to his own satisfaction, that morality is not susceptible of demonstration. He now proceeds to attempt to prove that it does not consist in any matter of fact. His aim, it will be remembered, is to show that, since morality is neither susceptible of demonstration, nor consists in any matter of fact, thought and the understanding are not capable of 'fixing the boundaries of right and wrong'.

Since this part of his argument is not very long, and is difficult to summarize, I shall quote it in full.

'Nor does this reasoning only prove, that morality consists not in any relations, that are the objects of science; but if examin'd, will prove with equal certainty, that it consists not in any *matter of fact*, which can be discover'd by the understanding. This is the *second* part of our argument; and if it can be made evident, we may conclude, that morality is not an object of reason. But can there be any difficulty in proving, that vice and virtue are not matters of fact, whose existence we can infer by reason? Take any action allow'd to be vicious: Wilful murder, for instance. Examine it in all lights, and see if you can find that matter of fact, or real existence, which you call *vice*. In which ever way you take it, you find only certain passions, motives, volitions and thoughts. There is no other matter of fact in the case. The vice entirely escapes you, as long as you consider the object. You never can find it, till you turn your reflexion into your own breast, and find a sentiment of disapprobation, which arises in you, towards this action. Here is a matter of fact; but 'tis the object of feeling, not of reason. It lies in yourself, not in the object. So that when you pronounce any action or character to be vicious, you mean nothing, but that from the constitution of your nature you have a feeling or sentiment of blame from the contemplation of it. Vice and virtue, therefore, may be compar'd to sounds, colours, heat and cold, which, according to modern philosophy, are not qualities in objects, but perceptions in the mind: And this discovery in morals, like that other in physics, is to be regarded as a considerable advancement of the speculative sciences; tho', like that too, it has little or no influence on practice. Nothing can be more real, or concern us more, than our own sentiments of pleasure and uneasiness; and if these be favourable to virtue, and unfavourable to vice, no more can be requisite to the regulation of our conduct and behaviour' (468–9).

Comments

(1) Hume has promised (463) to show that reason does not establish moral distinctions, in that these are not established by a process of probable inference, which consists in inferring matters of fact. He does not here try to prove this conclusion directly, but attempts to establish it *a fortiori*, by showing that moral distinctions do not *consist* in any matters of fact.

(2) If moral distinctions do not consist in any matters of fact,

the conclusion that they are not discovered by inference would certainly follow. The converse, however, is not true. Moral distinctions could consist in matters of fact, such as that I am in pain or that there is some red-looking object in the centre of my visual field, although they were not established (by me, at any rate) by a process of inference. We do not infer that we feel pain, and, though *others* may infer that we feel pain, it is seldom held that statements about moral distinctions are statements about what others feel (though it is sometimes held that they are statements about what we ourselves feel).

(3) Hume says 'Nor does this reasoning only prove, that morality consists not in any relations ... but ... that it consists not in any *matter of fact* ...' (468). What reasoning is he referring to when he says 'this reasoning'? The most natural assumption to make is that he is referring to the reasoning he makes use of in the immediately preceding discussion of parricide and incest. It is quite obvious, however, that nothing he says here has the slightest tendency to show that morality does not consist in any matter of fact. Even if Hume is right in thinking that all the *relations* possessed by human parricide or incest on the one hand, are also possessed by animal and vegetable 'parricide' or 'incest' on the other, it is quite obvious that there is a simply enormous number of *facts* about human incest and parricide which are not facts about animal and vegetable 'incest' and 'parricide'. I have already said as much as is necessary about this (pp. 58–9).

(4) Hume argues that if you consider an action, for example, wilful murder, you will not, if you examine only the action itself, find any matter of fact which constitutes its vice, but only certain passions, motives, volitions, and thoughts (468). This is mere assertion, and Hume is guilty of appealing more to rhetoric than to argument. However, it must be admitted that it is an assertion not entirely without some semblance of plausibility. A psychologist, one feels, might be able to establish all the facts about an action, and leave it an open question whether this action was virtuous or vicious. But then, a psychologist might establish all the facts about an action, and leave it an open question whether it was legal or illegal; it does not follow from this that the legality or illegality of an action is not a fact about the action, rather than a fact of a different kind from those which a psychologist could establish. Hume is a bit inclined to suppose that if you cannot discover something about an action by 'examining it in all lights' then there can be no matter of fact to be discovered. You would

not expect, however, to be able to discover whether or not an action was illegal (or even whether or not it was unpopular) by examining it in any kind of light, so long as you confine your attention to the action itself. Hume is certainly drawing attention to, or perhaps making use of, our feeling that there is some difference between the fact that an action is vicious and other facts about the same action, but that an action's being vicious is not a fact about the action is something which needs much more careful proof than he gives it.

(5) Hume now goes on to say (inconsistently) that we *do* find the vice of wilful murder if we stop considering the action itself, but turn our attention to what goes on in our own breasts. What we find there is the sentiment of disapprobation which this action excites (469). If Hume is suggesting that, since the vice of wilful murder is not to be found in the action itself, it *must* be found in the sentiments it arouses in us, then this suggestion is untrue. It does not follow from the fact that the vice of wilful murder is *not* in the action, that it *is* in our sentiments. It might, for all Hume has yet said to the contrary, lie in the fact that it was contrary to the social conventions governing the society to which the man who performed the murder belonged, or contrary to the will of God, to mention only two possibilities.

(6) It is extremely puzzling that Hume says that he is going to show that morality is *not* a matter of fact, but ends up concluding that it *is* a *matter of fact about our sentiments*. It could be, of course, that what he really means to establish is that it is not a matter of fact *intrinsic* to the action we describe as vicious. It does not seem, however, that he does anything much to establish even this narrower conclusion. The argument, for example, that morality cannot consist in any matter of fact, because these only move us to action with the co-operation of a passion, whereas morality moves us to action without the co-operation of a passion, would show, if it shows anything, that morality did not consist in *any* matter of fact about the action, whether this is an extrinsic fact or an intrinsic one.

(7) If it is Hume's view that morality, though it consists in a matter of fact, does not consist in a matter of fact which could be established by either demonstrative or probable reasoning, then this would accord well with morality consisting in facts about our feelings. It can be argued that we do not argue or infer that we feel approval or disapproval of actions; we know such facts without inference. Hume is quite wrong, however, when he suggests that

facts such as that I disapprove of wilful murder are the objects of feelings, not of reason. They are *not* objects of *reason*, in the sense that I can neither prove by some *a priori* deductive argument that I must be having feelings of such-and-such a sort, nor need infer my feelings from any constant conjunction among my impressions. But it does not follow from this that they *are* the objects of *feelings*. We do not *feel* that we feel disapproval of wilful murder. Disapprobation *is* a feeling, not the object of feeling.

(8) The view that 'when you pronounce any action or character to be vicious, you mean nothing, but that from the constitution of your nature you have a feeling or sentiment of blame from the contemplation of it' (469) is one of the few philosophical theories that have been conclusively refuted, though this was not done, so far as I know, until G. E. Moore wrote *Ethics* in 1912. I explain how it is refuted in a later chapter (pp. 112–13).

(9) Hume likens vice and virtue to secondary qualities like sounds, colours, heat and cold, which 'are not qualities in objects, but perceptions in the mind' (469). It does not seem to me that the resemblance between virtue and vice and these secondary qualities is very great.

The reason why primary qualities were distinguished from secondary qualities is this. Primary qualities, unlike secondary qualities, were held to be 'in' the object which was supposed to possess them. The reason for holding this was supposed to be that the secondary qualities, but not the primary qualities, were dependent upon the conditions under which the object was observed. For example, since a piece of material might look blue in artificial light, but green in daylight, it was argued that neither colour was 'in' the material. All that was in the material object was the power of causing men to have sensations of seeing blue in certain circumstances, green in others.

This distinction was later criticized (by Berkeley) on the grounds that the primary qualities of an object were as much dependent upon the sense organs of the observer and the conditions under which the object was observed as were its secondary qualities. Berkeley also thought that it was impossible for an object to possess primary qualities without also possessing some secondary qualities or other. It would take me too far afield, however, to discuss in detail the distinction between primary and secondary qualities here, though I believe the distinction to be a valid one. For even if you (mistakenly) identify the virtue and vice of an action with its power or disposition to produce feelings of approval or disapproval in (normal) observers, it is, I think, quite wrong to treat this disposition as a secondary quality. If we open our eyes, a thing's apparent colour is thereby revealed to us. If we listen, we hear the noise it makes, which noise will sound loud or soft according as to the state of our ears and brain, the distance it is from us, and so on. If we feel it with our hands or some other part of our bodies, it will feel rough or smooth. If we taste it, it will taste sweet or bitter. If we smell it, it will smell acrid or pungent. In all these cases, a sensation is immediately produced in us by physical contact with the object, in the case of touch, or by light waves emanating from the object falling on the retina, or by sound waves from the object falling on our ear drums, or by minute particles from the object acting on our nostrils. There is no such causal action upon our bodies in the case of our feelings of approval or disapproval. We can disapprove as readily of actions we do not witness as we can of those we do. The actions we approve of or disapprove of do not have to be contemporary with our approval; they may be past actions, or actions yet to be performed, or actions which never are performed, but are simply contemplated. What, in fact, causes us to feel approval or disapproval of an action is not, strictly speaking, the action itself, but our beliefs about it. In the case of the secondary qualities, the sensations come first, and our beliefs about the object are based on them. In the case of our feelings of approval or disapproval, our beliefs about the action come first, and our feelings of approval or disapproval are based upon these. We do not hear noises as readily from an object which causes no sound waves as we do from an object which does, but we do disapprove as readily of what we wrongly believe to be a case of unprovoked violence as we do of an action which we rightly believe to be a case of unprovoked violence. Indeed, the *fact* that an action is one of unprovoked violence will cause no disapproval in us, unless we know or truly believe that it is a fact.

  It does not, of course, at all follow from the fact that secondary qualities are dispositional, but primary qualities not dispositional, that objects do possess primary qualities, but do not possess secondary qualities. Hence, even if Hume were right in holding (here, at any rate) that being virtuous and being vicious are secondary qualities, it does not follow that actions are not virtuous or vicious, or, as has been suggested, that Hume thought that they were not virtuous or vicious.

  Though there is an important difference between an object's

disposition to look red and feel smooth, and the fact that our beliefs about actions or supposed actions arouse in us feelings of approval or disapproval, it does not follow from this that there is not also an important resemblance. One possible view of secondary qualities is that, though in fact all there is in the object is the power of producing sensations of red etc. in observers, ordinary people, when they say that an object is red, are attributing to it a non-dispositional property which in fact it does not have. We believe, for example, that redness is spread evenly over the surface of the cricket ball, both on the front, which we see, and on the back, which we do not see. If redness is not in the cricket ball, this belief is false. All that is true of the cricket ball is that it has the power to make us see red when our eyes are suitably situated relatively to it. It might, then, be held that the same is true of virtue and vice. All that there is in an action is its power to evoke in us feelings of approval. But ordinary people think (wrongly) that there is more to it than this. They, perhaps, project their feelings on to the actions, and, whereas all that is in fact the case is that the thought of the action arouses feelings of approval or disapproval in them, they think that the action has some quality in itself. What are in fact simply people's feelings get detached, so to speak, from the people who have them, and seem to cling inescapably to the actions which arouse them.

There are two possible variants of such a view. The first is that ordinary people are attributing some such projected quality as I have described to the actions which they judge to be virtuous or vicious. The second is that all they are doing when they say that actions are virtuous or vicious is attributing to actions the power to arouse feelings of approval or disapproval in themselves, but that they wrongly think they are doing something else, namely, attributing some inherent quality to the action. From the second of these two views, all that follows is that they have a mistaken theory about what they are doing when they make moral judgements. It does not follow that the moral judgements they make are mistaken. But from the first of these two views, it does follow that all ordinary people's moral judgements, to the effect that actions are virtuous and vicious, are mistaken, for it is simply not true that actions themselves possess anything other than (at most, as we have seen) the power to arouse feelings of approval or disapproval in the people contemplating them. (Though, on this first view, it will follow that all moral judgements ascribing virtue and vice to actions are false, some, nevertheless, will be more false than others. If we say of an action that does arouse disapproval that it is vicious, then what we say will be false, because the action has no inherent quality, but simply the disposition or power to arouse such feelings. Nevertheless, we can reduce or modify our original claim in such a way that it becomes true. If, on the other hand, we say that an action is vicious, and it does not even arouse feelings of approval, and, perhaps, even arouses feelings of disapproval, then what we say is totally and completely false. In this case, our claim cannot be reduced or modified in such a way as to become true.)

There is no reason at all, however, to think that Hume thought that the moral judgements of ordinary people were false. It would indeed, be quite extraordinary for a philosopher who thought that all judgements to the effect that certain actions were virtuous and others vicious were false to go on to give an account of what kinds of actions were virtuous and what kinds of action were vicious. It follows, therefore, that if Hume did think that ordinary people supposed that virtue and vice were 'in' the actions, when in fact they were not, that the only mistake he was attributing to these people was the mistake of holding an incorrect theory about the nature of virtue and vice, not the mistake of thinking that actions were virtuous and vicious when in fact they could never be either.

(10) Hume concludes the paragraph we have been discussing by saying that, if our sentiments are 'favourable to virtue, and unfavourable to vice, no more can be requisite to the regulation of our conduct and behaviour'. This is doubtless true, but why does he say it? I think that he may be drawing attention to the fact that, on his theory, there is no need to explain why virtue attracts people, and why they sometimes act virtuously, by explaining that virtue is really to their long-term interest, in this world or the next. Virtue, according to Hume, has an immediate appeal to us. Indeed, if what Hume says in *this* paragraph is right, the fact that virtue has an immediate appeal is an analytic proposition, not a synthetic one. Virtue is defined as whatever actions or characteristics have an immediate appeal, of a certain sort. Hume is then in a position to claim that he can do what he has argued rationalists cannot do, explain how it is that it is a necessary truth that our sentiments are favourable to virtue, and unfavourable to vice.

(11) In this paragraph Hume makes the claim that his view about the nature of virtue and vice will have little effect on practice (469). This, presumably, is because it is not a view about what things are in fact virtuous and what things are in fact vicious

(though Hume later puts forward such a view). This is another reason for thinking that he did not think that all judgements ascribing virtue and vice to actions are *false*, though one could argue that even this apparently radical view would have no effect on practice, because it would be psychologically impossible for anyone to take it seriously outside his study. It is possible to divide the problems of moral philosophy— though the division is not exhaustive—into two kinds. There are those which concern the nature of moral judgements, the manner, if any, in which we come to know them to be true, and the nature of the properties we ascribe to things when we make them. On the other hand, there are questions of substance about what things possess these properties. It might well be Hume's view—and, if it is Hume's view, it is a plausible one—that only the questions of substance, as opposed to the epistemological questions, have or ought to have an effect on practice. This would be true, of course, only if the epistemological questions did not have a bearing on the substantial ones. If the result of our moral epistemology was to cause us to conclude that we could not ever know or have any good reason for thinking that actions and men ever were virtuous or vicious, then such a conclusion ought to have an effect on our practice, even if it does not.

V

HUME'S FOURTH ARGUMENT: OF OUGHT AND IS (469–70)

Hume's View

THE most influential passage in those sections of the *Treatise* with which we are concerned is that which contains the argument, inserted almost as an afterthought, in the last paragraph of Book III, Part I, Section I. I shall quote this passage in full.

'I cannot forbear adding to these reasonings an observation, which may, perhaps, be found of some importance. In every system of morality, which I have hitherto met with, I have always remark'd, that the author proceeds for some time in the ordinary way of reasoning, and establishes the being of a God, or makes observations concerning human affairs; when of a sudden I am surpriz'd to find, that instead of the usual copulations of propositions, *is*, and *is not*, I meet with no proposition that is not connected with an *ought*, or an *ought not*. This change is imperceptible; but is, however, of the last consequence. For as this *ought*, or *ought not*, expresses some new relation or affirmation, 'tis necessary that it shou'd be observ'd and explain'd; and at the same time that a reason should be given, for what seems altogether inconceivable, how this new relation can be a deduction from others, which are entirely different from it. But as authors do not commonly use this precaution, I shall presume to recommend it to the readers; and am persuaded, that this small attention wou'd subvert all the vulgar systems of morality, and let us see, that the distinction of vice and virtue is not founded merely on the relations of objects, nor is perceiv'd by reason' (469–70).

Comments

(1) Hume says that this derivation of an *ought* from an *is* only seems inconceivable (469). For this reason, coupled with the fact that Hume does later give reasons, in the form of statements about what is the case, for conclusions about what ought to be done, some philosophers have maintained that Hume thought that the

derivation of an *ought* from an *is* was possible, but simply required explanation. That Hume himself thought that the derivation of an *ought* from an *is* was impossible is, besides being quite obvious, shown by the fact that he thought that pointing out that they involved such a derivation 'wou'd subvert all the vulgar systems of morality . . .' If he had thought that the drivation only seemed impossible, he would, surely, have merely concluded that pointing out that it was involved seemed to subvert the vulgar systems of morality.

The most that Hume's argument will prove, of course, is that it is possible to assert premises involving *is*, and deny any alleged conclusion involving *ought*, without contradicting oneself. Such premises, therefore, cannot entail such conclusions. There are, however, reasons other than deductively conclusive reasons. Hence the fact that Hume himself gives reasons of some sort for statements involving an *ought* does not at all show that he thought it was, after all, possible to deduce logically an *ought* from an *is*. The fact that Hume gives reasons for *ought* statements would be inconsistent with the view that it is not possible to give reasons of any sort for an *ought* statement. If he did think this, as, perhaps, he ought to have done, he would have been inconsistent.

(2) It is obviously impossible to derive statements asserting what a man ought to be doing from statements asserting what he is doing. This, however, does not show that it is impossible to derive an *ought* statement from any *is* statement; merely that it is impossible to derive a statement like 'ought to be Xing' from one like 'is Xing', where the symbol X is replaced by the same word in each case.

(3) For a 'formal' contradiction between accepting the premises of an argument and rejecting its conclusion to be discernible, it is necessary for the same words to occur in the contradictory of the conclusion (and so in the conclusion) and in the premises. Though such a contradiction can be seen to obtain between 'All men are mortal and Socrates is a man' and 'Socrates is not mortal', no such contradiction is discernible between 'All men are mortal and Socrates is a man' and 'Socrates will live for ever'. None of the usual tests for whether a syllogism is valid will tell you whether 'is not mortal' and 'will live for ever' do or do not mean the same thing. Such a formal contradiction can be detected only if the same words occur in the premiss or premises and the contradictory of the conclusion. Hence no formal contradiction will ever be discerned between premises which do not contain the word 'ought'

and the contradictory of a conclusion which does. This does not mean that there is no contradiction at all, because it *is* contradictory to maintain both that all men are mortal and Socrates is a man, and that Socrates will live for ever.

(4) Some philosophers have felt themselves compelled to admit that an *ought* is logically deducible from an *is*. Either 'Either today is Tuesday, or I ought to apologize' is an ethical proposition, or it is not. If it is, then it is entailed by 'Today is Tuesday', which is factual. If it is not, then it, being factual, together with the factual proposition 'Today is Tuesday' entails 'I ought to apologize'. There is a discernible stink of fish about this argument. A man who maintained that information about botany was irrelevant to astronomy would not be impressed by being told that 'Flowers are vegetables' entailed 'Either flowers are vegetables, or the moon is made of green cheese'. What, then is wrong with it?

It seems odd to classify 'Either today is Tuesday, or I ought to apologize' as either an ethical statement or a factual one. (It would be equally odd to classify 'Either flowers are vegetables, or the moon is made of green cheese' either as botanical or as astronomical.) When we are faced with compounds of two statements, one of which is factual and the other ethical, there seems to be no good reason why we should say that the compound itself is either one or the other. This means, of course, that the disjunction 'ethical or factual' is not exhaustive; compounds of both are at least sometimes neither. This allows us to say that when 'Either today is Tuesday or I ought to apologize' is entailed by 'Today is Tuesday', this is a case of a non-factual proposition being entailed by a factual one, not a case of an *ethical* proposition being entailed by a factual one (for we have agreed that the former proposition is neither factual nor ethical). Similarly, of the entailment of 'He ought to apologize' by 'Today is not Tuesday' and 'Either today is Tuesday, or he ought to apologize', we have to say that an ethical proposition is entailed by non-ethical propositions, though we do not have to say that an ethical proposition is entailed by factual ones.

(5) If 'I ought' implies 'I can', then 'I cannot' must imply 'It is not the case that I ought'. Hence a factual proposition *can* entail an ethical one. I am not sure, however, that 'I ought' does entail 'I can'. It may be that there are many things which I ought to do that I cannot, but that there is no point in saying to someone, in a disapproving manner, that he ought to have done something, if this is something he could not have done.

(6) I have already shown that no 'formal' contradiction can be demonstrated to hold between premisses framed in terms of *is* and the contradictory of an alleged conclusion framed in terms of *ought*. There is no reason, however, why an informal contradiction should not exist. 'Some bachelors are married' is obviously contradictory, but is not, like 'Some unmarried men are married', formally contradictory. Similarly, 'Jones knows *p*, but *p* is not true' involves no formal contradiction. There is nothing in the nature of the word 'ought' to prevent its being deducible from the word 'is' any more than there is something in the nature of the word 'bachelor' to prevent its being equivalent to the phrase 'unmarried man'. Both are recognized by their sound or shape, not by what they are or are not synonymous with.

(7) Where *some* so-called 'non-ethical' uses of 'ought' are concerned, I am strongly inclined to think that there is an 'informal' deductive passage from an *is* to an *ought*. Take the statement 'The train ought to have been here five minutes ago'. Is it necessary intuitively to apprehend a synthetic connection, between the time at which it was due and the time at which it ought to have arrived, or to establish inductively that trains ought to arrive when they are due, or to discern this truth by means of a special sense, which sees that trains which arrive when they are due have the peculiar sensible quality of arriving when they ought? Surely all I need in order to infer 'This train ought to have arrived five minutes ago' from 'This train was due five minutes ago' is to know how to use the words 'ought' and 'due', etc. To take another example, is anything else needed in order to infer 'That bridge ought not to have fallen down' from 'That bridge was made by a competent architect, by a reliable firm of contractors, out of excellent materials, and commissioned by the city corporation on the understanding that it would last for at least a hundred years'. Can we not pass to 'I ought not to have trumped my partner's Ace' from 'Trumping my partner's Ace was the only thing I could have done to prevent us winning a game which we both I and my partner wanted to win and were trying to win, and where there were no considerations (such as wanting to flatter our opponents) in favour of deliberately losing'? Where so-called 'moral' uses of 'ought' are concerned, the case is perhaps more difficult; 'good boy' would seem to follow fairly easily from 'boy who is quiet, tidy, polite, and obedient', and 'my duty' from 'task allotted to me by my lawful superior in the institution of which I am a member', but quite what 'morally wrong' follows from is more difficult to determine.

(8) Though it may be possible to build a bridge between matters of fact and morals, there is a sense in which it is not the sort of bridge-building which will enable us to solve any moral problems, as opposed to epistemological ones. The fact that there is no logical gap between the time at which the train is due and the time at which it ought to arrive will not in the least enable me to know what time it is due; similarly, if it were possible to show that there was no logical gap between a number of facts about promise-breaking and the moral judgement that promise-breaking is wrong, this would not in the least enable me to know whether promise-breaking is wrong, if I am in ignorance of these facts.

(9) Hume's actual words, in the passage to which we have given most attention are 'For as this *ought*, or *ought not*, expresses some new relation or affirmation, 'tis necessary that it shou'd be observ'd and explain'd; and at the same time that a reason should be given, for what seems altogether inconceivable, how this new relation can be a deduction from others, which are entirely different from it' (469). Of course, if it is a new relation, meaning one not already mentioned in the same or in other words, then no deduction is possible. Anyone disagreeing with Hume must do so by showing that it is not, as it seems, a new relation, but one which has already been mentioned, though not by use of the word 'ought'.

(10) Some philosophers (for example, Professor Searle in *Speech Acts*, Cambridge, 1969) have argued that it is possible to deduce *ought* statements from 'institutional' statements such as that someone has made a promise, owns property, or has put my King in check. These are institutional statements, because promising, property, and putting a King in check can only be understood if you understand that there are rules demanding that promises be kept, property be at least half sacred, and that, if my King is in check, I must prevent it from being taken.

Let us consider promising as the type or exemplar of the others. Searle argues that from the statement 'Smith says "I promise",' which is clearly factual, it follows, provided Smith is serious, in the presence of a promisee who is not deaf, etc., that Smith has made a promise. From the fact that Smith has made a promise it follows that Smith has put himself under an obligation to keep it, from which it follows that Smith is under an obligation to keep it, from which it follows that Smith ought to keep it.

I propose to take a short way with this argument, without considering in detail the steps from one provisional conclusion to the next. Either, when I say 'Smith promised to do *A*' I intend to imply that Smith ought to do A, or I do not. If I do intend to imply this, then that Smith ought to do A does follow from 'Smith promised to do A', but the statement that Smith promised to do A is not purely factual. On the other hand, if I do not intend to imply this, then the statement that Smith promised to do A is purely factual, but does not imply that Smith ought to do A. I ought not to describe people as having promised if I do not think that they ought to do what they have promised, but use some other form of words, such as 'Smith went through the motions of promising' or 'Smith seemed to promise' or 'Smith said "I promise"', but failed thereby to make a promise (since he did not bring about a situation in which he ought to do what he promised)'.

It is surely obvious that the statement that any given promise ought to be kept is a synthetic one. What are we to say, however, about the (true) statement that most promises ought to be kept? Is it synthetic, or is it analytic? For myself, I think it is synthetic, for I think that someone questioning the value of the practice or institution of promising might wish to hold that all promises ought to be broken in order to destroy this institution or this practice. I think it would be foolish to say that someone questioning the institution meant something different by 'promising' from what someone accepting the institution means; they are talking about exactly the same things when they talk about promising. Hence, if the proposition that most promises ought to be kept is synthetic but believed to be false by those questioning the institution, it must also be synthetic for those who accept the institution and believe this proposition to be true.

The fact/value gap: independent treatment of the problem

When we consider so-called 'non-moral' uses of ethical words, it looks implausible to suggest that an 'ought' cannot be derived from an 'is', and equally implausible, for that matter, to suggest that a 'right' or a 'wrong', or a 'good' or a 'bad', cannot be derived from an 'is'. When we consider bringing someone the screwdriver he has asked for (the right screwdriver), taking the road which leads to the place to which we wanted to get (the right road), making the move which brings about the desired defeat of our opponent (the right move), producing that figure which is the sum of a number of smaller figures we were asked to add (the right answer), guessing that the card in someone's hand is the Ace of Spades when it is (the right guess), doing something in a way which achieves what I want economically and efficiently (the right way), telling the time it is, which is what one wants to know (the right time), being at the place and time pre-arranged, and necessary to the fulfilment of some co-operative enterprise (the right place and time), choosing the man who will perform the task allocated to him successfully (choosing the right man), it is difficult to avoid the conclusion that all these things have something generic in common, which we use the English word 'right' to signify. There is no other word in English which does quite the same job, or has quite the same meaning, as the word 'right' in its 'non-moral' senses. However, though there is no synonymous English word, the word 'right' is closely allied to words and phrases like 'hit', 'success', 'achieving what we want', 'doing what we are asked', etc., and the word 'wrong' to their opposites.

In at least one important class of cases, therefore, we use the word 'right' of the successful way of doing something, and the successful way is the rational way for anyone who wants to do that thing to adopt. In this case, then, at any rate, the gap between 'is' and 'right' is fairly easy to close. If to say that something is the right way to do something is to say that it is the rational way to do it—it is clearly irrational to take the wrong road to Rome, if you want to get there—and if what is rational to do is determined by people's ends and the means to attaining them, then certain facts, about men's ends and how they may be realized, do entail that certain courses of action are rational or irrational. Thus they determine what, in this 'non-moral' sense, are right or wrong.

When we turn our attention to the passage from 'is' to 'good', the situation is much the same. A car that is fast, reliable, comfortable, and cheap to run, a knife that is sharp and durable, an answer that is clear, brief, comprehensive, and to the point, a house which is solid, commodious, pleasing, and sensibly planned, and a cricketer who is strong, fit, lynx-eyed, accurate, and fluent have something in common, which is that they are a good car, knife, house, or cricketer. It is not that they have some extra, mysterious characteristic, over and above the ones I have mentioned. Being efficient, reliable, comfortable, sharp, etc. is just the determinate form that their indeterminate goodness takes.

The passage from being efficient etc. to being good is therefore like the passage from being bottle-green to being green, and from being green to being coloured. And just as something can be coloured, without being green, so something can be good without being efficient.

There is, however, a covert anthropomorphic reference in the word 'good'. If men were only six inches high, they would not call good the houses which they now call good. Indeed, such houses would not be good houses, for they would be much too large for the people for whom they were intended. Because of this covert reference to the needs of human beings, it is impossible to discover the goodness of anything, though we 'turn in on all sides' as much as we like. This is possibly one reason why Hume failed to find it, and thought that it must be our feelings about things which determined their goodness. But a good car is a good car, whatever our attitude to it is. There is the possibility of confusion on the following point. In one way, the goodness of a car is not independent of our attitude to it. If we did not prefer speed for its own sake, or want to get to places quickly, and did not feel more at ease in leather upholstery, then having a fast engine and leather upholstery would not be characteristics tending to make the car which possessed them a good one. If we did not dislike work, or have other things to do with our time, being labour-saving would not be a characteristic tending to make good the house which possessed it. But given that we do feel more at ease in leather, and do dislike house-work, then cars or houses like this tend to be good, whether we think they tend to be good or not.

Saying that something is good also carries implications about what the speaker thinks it is rational to choose, as Professor Hare has pointed out. If I say that A is a good X, and B a bad X, it would be inconsistent, *ceteris paribus* (they may not be the same price, for example), to advise someone who wants an X to choose B rather than A, and irrational to choose B rather than A myself. If to say that something is good implies that it is rational to choose it (either for its own sake, or as an instrument for attaining something else), then it is possible to show that having certain characteristics logically entails being good. It is perhaps simpler to start by showing that if a thing is not good, it is not rational to choose it (again, *ceteris paribus*; a bad one might be the only one that I can find or afford). For example, if I want a golf ball, for the usual reason one wants a golf ball, it would be irrational to choose one which flew like a boomerang, or would not carry more than ten feet. Conversely, if it flies straight and far, it is a good golf ball, and it would, *ceteris paribus*, be rational to choose it. This enables us to show the logical entailment between the possession of certain factual properties and being good. For there are, in a golf ball, certain factual properties which would obviously make it irrational to choose it; hence, if it has these properties, it logically must be bad. If it has factual properties which would make it rational to choose it, then, logically, it must be good. In the case of golf balls, and many other things, the rationality of choosing things with such-and-such properties is derived from the end to the attainment of which they are an instrument. If your end is to get balls down holes, then it is irrational to choose balls which do not easily go down holes.

When it comes to the relation between 'is' and 'ought', in its 'non-moral' sense, it seems to me that it is also impossible—logically impossible—to accept certain factual premisses and deny that these premisses demand that such-and-such ought to be done. To take again an example which I have already used, it seems to me that, if anyone agrees that there is one and only one way for White to win a game of chess, he must agree that White ought to take that way, in order to win. If he disputes this, it must be because he simply does not know what the word 'ought' means. Of course, as we have seen, nothing about what White ought to do follows from any statement about what he is doing. It follows only from statements about what his aims are, statements about the rules of chess, and statements about what will or will not be the consequences of the various alternative moves these rules leave open to him. The word 'ought' draws attention to a comparison between what is done, or what does happen, and something that may or may not happen. It involves a comparison between the move White actually makes, and the move which he would have to make in order to achieve his ends, between the time the clock tells, and the time it would tell if it told the time it was, between what a man does, and what he would do if he did what he agreed to do, what he was told to do, what it was planned he should do, what he usually did, or what it could be reasonably predicted he would do. Though we cannot deduce an 'ought' when we know only one of the terms in the comparison—the move that *is* being made, for example, or the time the clock actually is telling—perhaps we can deduce it if we know both terms—the move that a man is making and the move that would result in his winning; the time the clock is telling and the time it is.

In an important kind of case, the word 'ought' also has implications about what it would be rational to choose. We do not say that someone ought to do something in a certain way unless we think it would be rational for him to do that thing in that way. We say that farmers ought to plant seed in spring, rather than in autumn, because we think it would be irrational for farmers to plant seed in autumn, and rational to plant it in spring. This is because farmers will not achieve the ends they have in view if they plant seed in autumn, but probably will if they plant seed in spring. If anyone doubts the rationality of this, it must be because he does not know what the word 'rational' means. Though the facts, together with the ends of farmers, determine what is rational for farmers to do, and so determine what farmers ought to do, we do not infer from the fact, by a logical argument, what it would be rational for them to do, or what they ought to do. It is rather an assessment of what it would be rational for farmers to do, and so of what farmers ought to do, based on the facts. This does not mean, however, that the rules for the use of the words 'ought' and 'rational' allow us to make any judgement we please, however eccentric, upon the facts.

When the word 'ought' is being used to tell someone what to do in order to achieve some end he is aiming at, there seems to be also the possibility of the following argument. If someone thinks that X ought to do Y, he is right in thinking this, and it really is the case that X ought to do Y, if X's doing Y is a good way, or the best way, of achieving what X is aiming at. As I have already said, being a good way of achieving what X is aiming at is just a generic characteristic which this way has, if it accomplishes what X wants cheaply, quickly, reliably, efficiently, and so on. It must logically follow from the fact that this way of achieving what X wants is quick, cheap, and efficient, etc., that it is a good way, for anyone who knew that it was a quick, cheap, and efficient way of achieving what X wanted, and doubted whether it was a good way, must do this because he does not know what the word 'good' means.

An alternative would be to argue that, if anyone says to X that he ought to do Y, his advice is good, sound, and sensible, and he *ought* to do Y, if what he is advised to do is a good way of achieving the end the means to which he was advised on. Again, if anyone is disposed to think that the means he was advised to take are a good way of achieving his end, but that the advice, nevertheless, is not good, sound, and sensible, he simply does not know what the word 'good' means.

Let us now turn to the so-called moral sense of the words 'right' and 'wrong'. We often say an action is right or wrong according as to whether it is permitted or prohibited by certain rules. If I am instructing someone in how to play chess, and start simply by teaching him the rules, and ignore strategy, I say 'That's wrong' when he makes a move which breaks one of them and 'That's (all) right' when he does not. When I am using the words 'right' and 'wrong' morally, there are, of course, no actual rules to which I am referring, for it may sometimes be (morally) right to break actual rules, and even (morally) wrong not to break them. The rules I have in mind are not any rules which are either followed or enforced, but rules which I am trying to get adopted. I am right in thinking that an action is right or wrong, and these actions *are* right and *are* wrong, if the rules which enjoin them are good rules. And, within certain broad limits, someone who knows what these rules are, and the situation to which they are applied, and the results of applying them, and fails to see that they are good or bad, as the case may be, simply does not know what the words 'good' and 'bad' mean. If the rules bring about man's ruin, they are clearly, to put it mildly, bad; if they make men happy, prosperous, and harmonious, they are, equally clearly, good. Rules which are good rules for society to have will be rules which are good rules for society to adopt, or retain, if they already are adopted.

The word 'good' in its so-called moral sense has very much the same kind of meaning it has when it is used in a non-moral sense. If a man has enough of virtues such as kindness, honesty, courage, temperance, industry, justice, and what might be described as 'moral-law abidingness', and has these characteristics in a sufficiently high degree, then he must be a good man, and anyone who doubts this does not know what the word 'good' means. These characteristics are virtues for pretty much the reason that Hume gives, that is, they are virtues because they are characteristics useful or agreeable to ourselves or others (though Hume probably exaggerates the morality, as opposed to the attractiveness and desirability, of the agreeable characteristics). It is very likely that we will have a favourable attitude to the possession of such characteristics, since egotism will cause us to be in favour of those virtues which are useful to ourselves, and humanity to be in favour of those which are useful to others. If we are highly egotistical and devoid of humanity, however, we could be in favour of ourselves possessing those virtues which are useful to ourselves, but not those which are useful to others, and in favour of others possessing those virtues which are useful to ourselves but not those which are useful to them. Though someone who is honest, kind, etc. cannot (logically) but be good, we do not establish first that someone is honest, kind, etc. and then infer that he is good. In establishing that he is honest, kind, etc. we are *ipso facto* establishing that he is good.

I think that there is again a connection between being a (morally) good man and what it is rational to choose. The connection cannot, though as in the case of 'ought' in its 'moral' sense it can, lie directly between being good and the ends of the individual man who is good, for a man may have to sacrifice some of his own ends in order to be morally good. The connection may lie, however, in the relationship between being morally good and being the kind of man it would be rational for society to want its members, in so far as their behaviour is a matter for public concern (which is an important qualification), to be. There is a very close logical connection between being morally good and being moral-law abiding, i.e. observing those rules which men ought to observe. If the rules which men ought to observe are rules which it is rational for society to have, then it must also be rational for society to want its members to be disposed to obey these rules. If which men are morally good is determined by what rules it is rational for society to adopt, and if the meaning of 'rational' is linked to the facts in such a way that it would be improper to describe as rational behaviour of just any kind we like, then what men are good men is determined by the facts, and it is logically impossible, given certain facts, for a man of such-and-such a kind to be anything other than a good man.

Where the so-called moral sense of 'ought' is concerned, the case is much the same as with the 'non-moral' sense. If I say that X ought to do Y, then there must be some rule from which it follows that X ought to do Y, and X *ought* to do Y, if there is some rule, which would enjoin upon X the doing of Y, and this rule is a good rule. Within certain broad limits, a description of the rule and the circumstances to which it applies and the results of applying it logically entail whether it is a good rule or not. If the results of applying a rule were disastrous for mankind, a person who called it a good rule would simply not know what the word 'good' meant. If the results of applying a rule were to cause men to be happy, healthy, prosperous, and to live harmoniously with one another, someone not calling it a good rule would simply not know what the word 'good' meant. And, *pace* Hume and many others, such a rule would continue to be a good rule, whether men supposed it to be a good rule or not, and whatever feelings it might, if they were sufficiently extraordinary, arouse in them.

We have seen that, when we are advising someone on the correct means to achieving a given end, there is a close connection between what we think he ought to do and what we think it would be rational for him to do. The connection cannot be quite the same when we say that someone (morally) ought to do something, for we might think that someone morally ought to do something though none of his own personal ends were realized by his doing it, and even if they were sacrificed by his doing it. The only suggestion I can make is that what a man morally ought to do has some connection with what rules it would be rational for society to have. I suggest that a man morally ought to do something, if there is some rule which would enjoin upon him the doing of that thing, and if it would be rational for society to adopt this rule, to the extent of applying the pressure involved in using the expression 'You ought to do that' to get people to obey the rule. Since it would clearly be irrational for society to allow indiscriminate killing, and still more irrational for it to enjoin indiscriminate killing, the gap between *is* and *morally ought* is closed in the following way. It is closed by showing that certain facts about rules, and about the situations to which they are applied, the consequences of applying them, and the ends people have, entail that certain rules are rational for society to adopt. They therefore entail that people morally ought to obey these rules.

One difficulty with what I have been rather sketchily arguing for in the last few paragraphs is that I give an account of the rationality of adopting various means (in terms of the effectiveness with which they bring about certain ends) but give no account of the rationality of seeking one end rather than another. An action can be rational, one would have thought, only if it is a rational way of achieving a rational end; to be a rational way of achieving just any sort of end is not good enough.

There does, however, seem to me something which makes some ends more rational ones to pursue than others. Characteristics which tend to make it rational to pursue an end are such things as the fact that doing so incidentally furthers other ends of the agent's, that it does not interfere with other ends he or other people also have, that it gives him some lasting satisfaction, and that there is a reasonable chance of securing it. It would be a mistake to say that all we aim at is pleasure (Hume pointed this out in Appendix II of the *Enquiry*) and that the 'ends' I have been talking about are simply means to pleasure. Many of our ends, if not most, are not sought for the pleasure they give us. Rather, we seek them for their own sake, and, in consequence, get pleasure when we attain them.

So far I have been speaking of what ends it is rational for individuals to adopt. But, if we define morality in terms of what is rational for society, are we not faced with the same problem when it comes to society's ends? Cannot society, as well as individuals, have rational as well as irrational ends?

Oddly enough, I do not think it can. An individual can have any ends he pleases, but society is not an individual and its ends are determined, logically, by the ends of the individuals who compose it. Hence, whatever ends individuals have, society's end must, logically, be to enable individuals to secure their ends, in a harmonious way, and to distribute justly, economically, and fruitfully the means to attaining them. If it apparently seeks any ends other than these, it is not seeking society's ends, but the imaginary ends of some fictitious aggregate, and this only to disguise the fact that it is being used, consciously or otherwise, as a mask behind which a powerful minority can further their own ends at the expense of other people's.

VI

ARGUMENTS FROM THE ENQUIRY

THE emphasis of *An Enquiry concerning the Principles of Morals* is on elegance, clarity, and persuasiveness. It was published in 1751, eleven years after Book III of *A Treatise of Human Nature* met with an indifferent reception, and Hume, presumably, was anxious to gain a hearing for his views which the more rugged earlier book did not at first secure. The pages on moral epistemology do not add a great deal to what Hume has already said in the *Treatise*. Indeed, a great deal is omitted. However, they do add something, and a discussion of Hume's views would be incomplete without a treatment of what he says in the later work.

For the purposes of this exposition, it will be best to start by considering Section I of the *Enquiry* together with Appendix I, pp. 285–7. I shall then discuss, or sometimes only mention, one by one, those of the five arguments against ethical rationalism not already dealt with, which Hume puts forward in the remainder of the first Appendix to the *Enquiry*.

Section I and Appendix I, pp. 285–7: the place of reason and sentiment

Hume starts Section I with an attack on those who deny the reality of moral distinctions. He attempts to refute them by saying that no one is so insensible as not to be 'touched with the images of Right and Wrong' or so prejudiced as not to observe that others are susceptible of like impressions (E170). The sceptic is, therefore, best left to himself; for, 'finding that nobody keeps up the controversy with him, it is probable he will, at last, of himself, from mere weariness, come over to the side of common sense and reason'.

Hume next turns to the controversy between those who think that knowledge of morals is obtained by 'a chain of argument and induction' and those who think it is given 'by an immediate feeling and finer internal sense' (E170). (In the *Treatise*, of course, he sided with those who held the latter view.)

In favour of the view that morality is discerned by reason, he adduces the facts that people *do* dispute about morals, put forward proofs and detect fallacies, cite examples, point out analogies, appeal to authority, and draw inferences from principles (E171). 'No man reasons concerning another's beauty; but frequently concerning the justice or injustice of his actions' (E171). And how, he asks, may we suppose that a different faculty is used to show that the prisoner did not do what he is accused of, and that, if he did do it, what he did was innocent? (E171.)

In favour of the view that morality is discerned by sentiment, he adduces the fact that virtue must be amiable; that something is loved cannot be proved by *a priori* argumentation (E171–2). The object of moral speculation is to make us eschew vice and embrace virtue, which would be impossible if what was honourable, fair, becoming, noble, or generous did not have an appeal to sentiment. Such an appeal is not possessed by what is evident, probable, or true, which latter procures only the cool assent of the speculative understanding (E172).

Hume's conclusion, put briefly, is that both sides are right, up to a point (E172). Reason and sentiment concur in moral determinations. The function of the former is to establish the facts of any matter (E173), upon which, when these are ascertained, sentiment subsequently passes judgement (E172).

The rest of the passages with which we are at the moment concerned are taken up with an account of what Hume believes to be the correct procedure for determining 'that complication of mental qualities, which ... we call Personal Merit' (E173). He again says that the way to determine this question is to discover what qualities are in fact regarded by men as estimable or odious, and to determine what it is that is common to each of these two classes (E174). He remarks, 'As this is a question of fact, not of abstract science, we can only expect success, by following the experimental method, and deducing general maxims from a comparison of particular instances' (E174). Men should be cured of their passions for 'hypotheses and systems' in moral disquisitions, as they are now cured in natural philosophy, and 'reject every system of ethics, however subtle or ingenious, which is not founded on fact and observation' (E175).

Comments

(1) Hume's refutation of moral scepticism is not very convincing. Moral sceptics would not deny that men supposed that there was a difference between virtue and vice, nor that they felt dif-

ferently about the two. They would argue that this difference in people's attitudes was the result of some naturally or artificially produced delusion. Hume's refutation of moral scepticism only works well on his own view, or one of them, that virtues and vices are simply those qualities of character which evoke universal approbation or the reverse.

(2) What Hume says on these pages is not in outline inconsistent with what he says in the *Treatise*. It is quite clear there that he does think that it is one function of reason to establish matters of fact, and that he does not wish to deny this when he says that moral distinctions are not apprehended by reason, but by a moral sense. He is simply being more conciliatory in the *Enquiry* than in the *Treatise*.

(3) Hume ought to have been a bit embarrassed, however, by an inconsistency on a point of detail, which I shall have occasion to mention again later (pp. 110–25 *passim*), between the *Treatise* and the *Enquiry*. For in the *Treatise* he does quite explicitly maintain (a) that morality does not consist in any matter of fact, capable of being discerned by the understanding, (b) that, in consequence, it cannot be established even by probable reasoning, and (c) that it is only because morality does not consist in any matter of fact that we can explain why apprehending a moral distinction can *alone* move us to action. In the *Enquiry*, however, he maintains that morality is a matter of fact, a fact about people's sentiments, and that it is established by observation of the sentiments of mankind. It is difficult to see how knowledge of facts about what people approve of or disapprove of can *alone* move us to action, any more than can knowledge of any other kind of fact.

(4) As in the *Treatise*, Hume is inclined to argue that, if morality were discerned by reason, rather than by sentiment, we would be indifferent to it. This is a very bad argument. At any rate, it needs a great deal more supplementation and elaboration than Hume gives it. It is quite easy, as I have already said, for a rationalist to explain how it is that we are not indifferent to morality, so long as he postulates, as seems entirely reasonable, a favourable attitude to morality, e.g. a desire to do what is right. Similarly, it is not difficult to explain how we are not indifferent to what the weather is going to be, given that we like it to be dry and warm, and hate it to be cold and wet. There is no need to postulate a special meteorological sense to detect what the weather is like, on the grounds that if we do not have this sense, our beliefs about the weather will not affect our behaviour.

What Hume ought to have done is to provide a detailed account of the manner in which our favourable attitude to morality differs from our favourable attitude to a large class of other things that can be analysed into the following two quite separate components: (a) a disposition to be in favour of anything we believe to possess a certain property, or be of a certain character, and (b) a belief that something possesses this property or is of this character. In these cases, it is just a contingent fact that we have a favourable attitude to things about which we have the belief. It just so happens that we are in favour of those things which we believe, say, to be low in nicotine or free from cholesterol.

There are, I think, two ways in which the relation between our beliefs about something, and our attitude to it, might differ from this. It might be—and this might be Hume's view—that it is a necessary truth that we are in favour of morality. It is true that Hume here states that morality is an empirical science, which determines what actions we have a morally favourable attitude to, but this presumably means that it is an empirical matter what are the *other* features of characteristics we have a morally favourable attitude to (according to Hume, that they are useful and agreeable), not that it is an empirical matter that we are morally in favour of those characteristics which are virtues.

The second way in which our attitude to morality might differ from these very common, if not standard, cases is this. Where, for example, being free from nicotine or cholesterol is concerned, it is perfectly possible to find out whether something is free from nicotine or cholesterol, regardless of our attitude to possessing these qualities. Discovering whether something is free from nicotine or cholesterol comes first, and our attitude to it results from this discovery. There *may* be cases, however, when it is simply impossible to find out whether or not something possesses a certain property—or, at least, impossible to find this out for oneself, without accepting the testimony of others—unless one's attitude to possessing the property is not neutral. It might, for example, be impossible to tell whether or not something was frightening unless you yourself felt frightened by it. It might be impossible to tell that something was disgusting, unless you yourself felt disgusted by it. It might be impossible to tell that a woman was beautiful, that a dress was elegant, that a tune was sad, or that a dance was graceful, or an action noble unless one had an attitude of a certain kind to these things, and was not indifferent to them.

The difference between these two different ways in which

beliefs may be connected with attitudes is that, in the first case, it is impossible for something to possess a property unless some attitude is taken up to it, whereas in the second case it is impossible only for anyone who does not have the attitude to discover whether anything possesses a certain property.

But why should this be so? I can think of the following possible explanations: (a) The belief is simply a belief about the attitude. In this case, we could not know the belief to be true, unless we had the attitude, because the belief would not *be* true unless we had the attitude. Such a view would explain why we could not be indifferent to morality, if to describe a character as virtuous simply meant that we approved of it. There is little to be said for such a view, however, for reasons which I shall give later (pp. 112–113). (b) The 'belief' might not be a genuine belief, but what is sometimes described as the expression of an emotion, rather than some description of a state of affairs towards which an attitude is taken up. In this case, there would be nothing for those who did not have the emotion or attitude to discover. For example, if I say 'Boo' to a goose, I am not describing the goose in any way, and so not describing the goose as having a feature which it is possible for those who do not share my attitude to geese to discover, or fail to discover. (c) It may be that I could not understand the words in the sentence which formulates the belief if I did not have the attitude. For example, if I never find anything odious, it may be that I simply do not know what the word 'odious' means. In this case, if I did not have the attitude, I might be able to discover what things people said were odious, but I would not know that anything was odious, for the sentence containing this word would be, for me, meaningless. (d) It might be that sometimes it was improper to describe something by a given word, unless you took up an attitude to the property connoted by the word. In such a case, one could find out whether or not that which was so described possessed the property in question, but could not appropriately use this word to say that it possessed it. To say this, one would have to find some other, neutral, word with the same connotation. For example, one might describe Smith as a Negro, whatever one's attitude to Negroes is, but one ought not (linguistically) to describe him as a nigger unless one dislikes Negroes. (e) It could be that, when we describe something as, say, disgusting, we are wrongly supposing it to have some characteristic of being inherently and essentially disgusting. Its disgustingness seems to cling inescapably to those things which are disgusting, to be an essential part of them, as their shape and size. Obviously, however, all that can be true of disgusting things is that they have the power of causing people to feel disgust. Their disgustingness is simply a false projection of these feelings on to the object. Hence the reason why it is impossible for anyone who does not find these things disgusting to discover whether the belief (that these things are disgusting) which arouses the disgust is true or not is because all such beliefs are simply false.

(5) In these passages, as in the equivalent passages in the *Treatise*, Hume fails to distinguish the platitude that if you 'extinguish all the warm feelings and prepossessions in favour of virtue, and all disgust or aversion to vice ... morality is no longer a practical study ...' (E172) from the philosophically important contention that our apprehension of morality moves us to action in a way different from the way in which we are normally moved to action, viz. by beliefs in conjunction with desires.

(6) Hume is similarly confused in the early part of the first Appendix to the *Enquiry*. He points out, correctly, that reason is necessary in order to instruct us in the pernicious or useful tendency of actions, especially in those cases of justice which are individually harmful, and performed only because it is generally useful to observe the rule which enjoins them. He then argues, also correctly, that, without a sentiment, this end would be indifferent to us. Here he says that the sentiment in question is humanity, which is a concern for the happiness of mankind. But all Hume can deduce from these premisses is that, without humanity, we would not do what tends to promote the happiness of others (or not, at any rate, without some ulterior motive). He cannot deduce, from these premisses, that, without humanity, we would not recognize the rectitude of seeking the happiness of others. His remark (E286) '*humanity* makes a distinction in favour of those which are useful and beneficial' is ambiguous. It may mean that, without humanity, we would have no preference for what is beneficial, or it may mean that humanity either produces or recognizes the morality of doing what is beneficial. Hume has produced no reasons whatsoever for thinking the latter to be true.

Hume's five arguments from the first Appendix to the Enquiry: *(1) Morality consists neither in a relation, nor in a matter of fact* (E287–9)

This argument adds nothing to what Hume has already said in the *Treatise*. Hence I shall not comment on it.

These pages, however, do contain an uncompromising statement of what we have already seen to be one of Hume's most characteristic views, though it is not a view to which he adhered consistently. On page 289 of the *Enquiry* he says 'The hypothesis which we embrace is plain. It maintains that morality is determined by sentiment. It defines virtue to be *whatever mental action or quality gives to a spectator the pleasing sentiment of approbation*; and vice the contrary. We then proceed to examine a plain matter of fact, to wit, what actions have this influence. We consider all the circumstances in which these actions agree, and thence endeavour to extract some general observations with regard to these sentiments.'

(2) We must be acquainted with all the facts about an action, before we pronounce upon its morality (E289–91)

Hume argues that 'a speculative reasoner concerning triangles or circles considers the several known and given relations of the parts of these figures, and thence *infers* [my italics] some unknown relation, which is dependent on the former. But in moral deliberations we must be acquainted beforehand with all the objects, and all their relations to each other; and from a comparison of the whole, fix our choice or approbation. No new fact to be ascertained; no new relation to be discovered.' Hume goes on 'The approbation or blame which then ensues, cannot be the work of the judgement, but of the heart; and is not a speculative proposition or affirmation, but an active feeling or sentiment. . . . In moral decisions, all the circumstances and relations must be previously known; and the mind, from the contemplation of the whole, feels some new impression of affection or disgust, esteem or contempt, approbation or blame' (E289–90).

Hume thinks that this shows why mistakes of right are commonly criminal, while mistakes of fact are not. When Oedipus killed Laius, he simply did not know that Laius was his father. 'But when Nero killed Agrippina, . . . all the circumstances of the fact, were previously known to him; but the motive of revenge, or fear, or interest, prevailed in his savage heart over the sentiments of duty and humanity. And when we express that detestation against him to which he himself, in a little time, became insensible, it is not that we see any relations, of which he was ignorant; but that, from the rectitude of our disposition, we feel sentiments against which he was hardened from flattery and a long perseverence in the most enormous crimes' (E290–1). Hume seems to conclude from this that moral determinations simply *consist* in the sentiments. When all the facts are known, 'nothing remains but to feel, on our part, some sentiment of blame or approbation; *whence* [my italics] we pronounce the action criminal or virtuous' (291).

Comments

(1) It is not, of course, necessary to know every fact about an action in order to pronounce upon its morality; indeed, this would be impossible. All that is necessary is that we should know all the relevant matters of fact (and also not be mistaken about any matter of fact). This qualification, however, does not affect Hume's argument in any way.

(2) Hume's statement that all the facts about an action must be known, before we pronounce upon its morality, might be regarded as question-begging. If that an action is wrong is a fact about it, then, if we do not know whether it is wrong or not, we do *not* know all the facts about it. There is, of course, a perfectly good sense of 'fact' in which we can know all the facts about an action, without knowing whether it is right or wrong. This is the sense in which 'question of fact' is opposed to 'question of right'. But the fact that we can know all the facts, in this sense of 'fact', about an action, without knowing whether it is right or wrong, does not show that there is not another, wider, sense of 'fact' in which that an action is right or wrong is a fact about it. This is the sense of 'fact' in which anything that is the case or is so is a fact, for it can be the case that an action is right or that it is wrong.

(3) Hume is quite correctly drawing our attention to the fact that we cannot know whether or not an action is right or wrong until we first know all the (relevant) non-moral facts about it, and also to the fact that, if we know all the (relevant) non-moral facts about it, this is all we need to know in order to know whether it is right or wrong. The morality of an action results from or depends upon the non-moral facts about it. Similarly, he points out, the fact that the angles of a figure add up to 180 degrees results from the fact that it is a triangle. Hume, however, thinks that we can *infer* that the angles of a figure add up to 180 degrees from the fact that it is a triangle, but we cannot *infer* from the non-moral facts about an action that it is wrong.

He is wrong to conclude merely from this, however, that whether or not an action is right or wrong is a matter of how we feel about it. We do not infer that a man is shabbily dressed from a detailed description of his attire. We perceive outright that he is shabbily dressed. This does not mean, however, that, whether or not a man is shabbily dressed is a matter of how we feel about his attire. Pretty obviously, it is not.

It may be that, in the case of being shabbily dressed, there is some detailed description of a man's attire from which it could be inferred that he was dressed shabbily. Being shabbily dressed is in this case a resultant attribute, like being morally wrong, which could (in the case of being shabbily dressed, though not necessarily in the case of being morally wrong) be inferred from a complete detailed description of his attire. But though this could be inferred, it is very doubtful whether anyone would in fact be able so to infer it. The analogy between being shabbily dressed and being morally wrong, however, breaks down. We do not need to know all the (other) facts about a man's attire in order to know whether or not he is shabbily dressed, even though whether or not a man is shabbily dressed is determined by these facts, but we do need to know all the (other) facts about a man's actions before we can tell whether or not it is wrong.

(4) In one sense of 'fact', we need to know all the facts about an action before we can decide whether or not it is illegal. This is the sense of 'fact' in which it is customary to contrast questions of law with questions of fact. It does not follow from this, however, that questions of law are determined by how people feel. Obviously, again, they are not.

(5) It is also not possible to *infer* from the facts about an action whether it is illegal or not. Again, it does not follow that whether or not an action is illegal is a question of how we feel. In order to know whether or not an action is illegal, we need to know all the facts about the action, and also to know what the law is.

Analogously, of course, one could say that, in order to know whether an action was right or wrong, one needed to know the facts about the action, and what the moral law is. In the case of the positive law of a given country, however, there are perfectly good procedures, and empirical procedures at that, for determining what the positive law is, and for specifying it, when it is ambiguous. In the case of the moral law, however, there are no such procedures. There are no statute books in which the moral law may be looked up, no case books of judges' decisions which may be appealed to. A judge, by deciding a case one way rather than another, may determine, to some extent, what the law is, but it would be absurd to suppose that any individual, by saying that an action was right rather than that it was wrong, could determine that, in future, actions of this sort should be right rather than wrong.

(7) It does not follow from this, however, that what rules a society should have is determined by how people feel. How people in fact feel is, as Hume frequently points out, a contingent matter, which could be otherwise. If it so happened that man's sentiment of humanity were replaced by one of universal malevolence, and men's moral sentiments were roughly the reverse of what they are now (E226), society would be unworkable, and mankind would perish miserably. Would it not be absurd to suppose that if men approved of actions which brought about man's downfall, this would mean that these actions were right? If this were to happen, would it not be possible to say that it was a bad thing that men approved of the actions they did, that it would be better if they did not approve of them, that they were wrong to approve of them, and, consequently, that they approved of actions which were not right?

(8) It seems to me that it would be possible for a being, looking at human society from outside, and who had no feelings of approval, or any other feelings, to determine what society's rules were, and subsequently to pronounce that some of these were good ones, others bad. He would not, of course, *care* what rules society had, or whether or not they were observed.

If, for some reason, he did wish to try to cause men to approve of rules which he thought were better for their society than the ones they had, there are two things he could do. He could tell people in this society not to bother about what was right and wrong, or he could use the words 'right' and 'wrong' in an attempt to get them to approve of rules which were more conducive to their interests. Such a being would, at least in part, know how to use the words 'right' and 'wrong', for he would use it with a view to causing people to approve of certain classes of action, regardless of whether these actions were to their tastes or inclinations, and with a view to causing them to perform those actions which he said were right, and to refrain from performing those actions he said were wrong. Unlike members of this society, of course, he himself would feel no disapproval of those actions which he said were wrong, so it might be that he was using these words improperly. It is arguable, however, that someone using 'wrong' of actions towards which he feels no disapproval is not misusing this word. We must not suppose that because we can infer, with a high degree of probability, that anyone who describes an action as wrong feels disapproval of this action, that he is using the word 'wrong' incorrectly if he does not disapprove of it. And of course, if someone says that something is wrong, we regard his own feelings as entirely irrelevant to whether what he says is true, or whether he is right in thinking this thing is wrong.

When this being disagrees with members of the society about what is right or wrong, is it possible to say whether he is right? Can you even say that this being is right or wrong, when, perhaps he says that actions are right or wrong not because he thinks they are right or wrong, but simply in order to cause people to perform or omit them? On the other hand, can you say that the people who disagree with this being are right, when their only reason for describing things as being right may be that they feel approval of them, since, as we have seen, 'right' does not simply mean 'approved of'? I can see nothing for it but to choose between them according as to whether the rules each are in favour of are good rules or bad rules.

It is fairly easy to tell, within certain limits, that some rules are good rules and others bad. For example, a rule enforcing homicide would be a bad one, and a rule prohibiting it a good one. Such rules would be good or bad quite irrespective of people's attitude to their infringement. This is not, of course, to say that such rules have nothing to do with people's wants. What rules are good rules must have a lot to do with what people want or need. (For example, they both want and need to be secure in the possession of their lives.)

Given the facts about certain rules, that they are good rules or bad is a matter of assessment of them. The passage from the facts to an assessment of the rules is not one of inference. Nevertheless, it is not a matter of how we feel, any more than what is a good move or a bad one at chess is a matter of how we feel. Hume might just as well have argued that since, in chess, a complete description of a move would leave out whether it was a good one or a bad one, whether a move *is* a good one or not will be determind by people's sentiments. Again, this is quite obviously untrue.

(9) Hume takes it for granted that when Nero's sentiments differed from those which, it is to be hoped, we would have felt in a like situation, this difference meant that he did not know what was right, though we would have done. It is, of course, possible, and perhaps more plausible, to argue that Nero did know what

was right, but, unlike ourselves, did not care. As I have said, a great deal of difficult argumentation is necessary to show that we cannot know what is right unless we care about doing it.

(10) It is not easy to see why Hume thinks that he has explained how it is that mistakes of right are commonly criminal, whereas mistakes of fact are commonly not criminal (if, indeed, it is true at all that this is so). It looks as if Hume supposes that to make a mistake of fact is to believe falsely that something has a property which it does not have, or lacks a property which it has. To make a mistake of right, on the other hand, is simply not to feel sentiments which most men would feel; or which it would be appropriate to feel; for example, most men would, quite properly, feel revulsion towards matricide, but Nero did not.

It is fairly easy to see that making a mistake on a matter of fact is not reprehensible, provided it is not the result of carelessness, negligence, laziness, conceit, or what have you. Such are the limitations of man's intellect that it seems inevitable that, *pace* Descartes, he will make many mistakes, however hard he tries to avoid doing so. It is not quite so easy to see, however, why it is reprehensible not to feel sentiments which other men feel, or to feel sentiments that other men do not feel. Quite often—as in the case of Nero—it may be true that our sentiments would have been different had we not deliberately allowed ourselves to become callous. There is, however, no reason why our deficiencies of sentiment should not be as frequently involuntary as our deficiencies of intellect.

However—as Hume himself points out in another part of the *Enquiry* (E313)—it is simply not the case that we disapprove only of what is voluntary. If we disapprove, as Hume says we do, of all traits of character which are harmful or disagreeable to ourselves or others, then we will disapprove of failure to feel such disapprobation; failure to feel disapprobation of harmful characteristics will itself be a (second order) harmful characteristic. Making a mistake of fact on any given occasion is not a characteristic, or disposition of character. We do, however, disapprove of character traits liable to result in their possessor making mistakes of fact, though this disapproval may, as in the case of stupidity, be tinged with pity, if the character trait is involuntary and harmful to its possessor as well as to others.

It may be that all deficiencies of moral sentiment are harmful, because it is always harmful not to approve of what is useful or agreeable; but then actual mistakes on matters of fact, and the dispositions which result in such mistakes, are also always liable to be harmful.

The best I can do for Hume is to suggest that to expect men never to make mistakes of fact would be unreasonable, whereas it would not be unreasonable to expect men always, or almost always, to feel appropriate sentiments. However, I am not at all sure that it is not unreasonable always to expect men to feel appropriate sentiments.

It looks as if Hume has chosen examples—the killing of Laius by Oedipus and of Agrippina by Nero—which are especially favourable to his case. Had he chosen to compare an incompetent doctor with a man wholeheartedly devoted to a noble but misguided cause, one might, from this latter example alone, have been inclined to take exactly the opposite view to that of Hume, and conclude that the doctor's mistakes of fact were more reprehensible than the other's misguided devotion.

(3) Morality not a quality of actions (E291-3)

'Euclid has fully explained all the qualities of the circle; but has not in any proposition said a word of its beauty. The reason is evident. The beauty is not a quality of the circle. ... It is only the effect which that figure produces upon the mind ... The orator may paint rage, insolence, barbarity on the one side; meekness, suffering, sorrow, innocence on the other. But if you feel no indignation or compassion arise in you from this complication of circumstances, you would in vain ask him, in what consists the crime or villainy, which he so vehemently exclaims against?' (E291-2.)

Comments

(1) It is difficult to talk much sense about 'beauty', because of the adulatory fog with which it is sometimes surrounded.

It is perhaps partly as a remedy for this that Austin recommended philosophers to turn their attention from 'the beautiful' to 'the dumpy'. A scientific description of a woman, limb by limb, organ by organ, nerve fibre by nerve fibre, molecule by molecule, would leave out the fact that she was beautiful; so it would leave out the fact that she was dumpy. Such a description of a man would leave out the fact that he was handsome (or ugly), or of a building that it was squat. However, this may not be because beautiful, dumpy, handsome, ugly, and squat are not properties of the things they appear to describe, but because they are descriptions belonging to what might be called a 'lower level of determinacy' than the 'more scientific' descriptions. A complete description of a building, wall by wall, window by window, brick by brick, may leave out the fact that it is squat, because being squat is an indeterminate description of the building as a whole, made unnecessary by a precise account of the nature and interrelations of its component parts. The same is certainly true of 'dumpy', and may be true of 'handsome' and 'beautiful'. If so, Hume's argument completely breaks down. That these are not characteristics of what they appear to describe is not shown by the fact that such descriptions are omitted from the more detailed inventory.

Perhaps the same thing could be said about being virtuous and good (p. 75). A detailed factual account of a man's dispositions, skills, capabilities, ambitions, and behaviour may mention that he is honest, kind, reliable, does not take what is not his, and plays a useful part in society. If it does not mention that he is good, perhaps this is because, after having enumerated these things, there is no need to add that the man is also good, any more than, having described the woman as four foot seven inches in height and three foot six inches in diameter, there is any need to add that she is dumpy. It is not that a woman has these measurements and is also dumpy, so much as that, in having these measurements, she is dumpy. Similarly, a man has not the aforementioned characteristics, and goodness as well; he has these characteristics, and is consequently good; in having these characteristics he is good. The relation between the measurements and being dumpy, and the relation between having the character traits and being good, is not such that we infer the latter from the former; rather, we recognize that having the determinate measurements is one way, though not the only way, of being dumpy, and having the determinate characteristics is one way, though not the only way, of being good. (There is no similar passage from being dumpy to having the measurements, or from being good to the characteristics, for there are more ways than one of being dumpy, just as there are more ways than one of being good.)

Even if this is true, so far as it goes, it cannot, however, be the whole story. In different circumstances from those which actually obtain in human life, kindness, honesty, etc. might not be good. Honesty, for example, is no great virtue in a spy. If human beings were very different from what they are, these characteristics might not be useful to mankind, and so be neither valuable nor valued.

But it is their usefulness that makes them good, not the fact that we value them, nor the fact, if it is a different fact, that we have favourable sentiments towards their possession. Of course, given that they are useful, it is only to be expected that we shall view them favourably; but they would be good, whether we viewed them favourably or not. It would be a great misfortune if men viewed good characteristics unfavourably, but this misfortune would consist in our viewing unfavourably characteristics which were good, not that these characteristics were in fact not good. There would, of course, be no one to say that they were good, but this is another matter.

(4) *All the relations possessed by actions may be possessed by inanimate objects* (E293)

This argument adds nothing to what has already been discussed (pp. 48–52).

(5) *No reason can be given why we desire morality* (E293–4)

Hume asserts, 'It appears evident that the ultimate ends of human actions can never, in any case, be accounted for by *reason*, but recommend themselves entirely to the sentiments and affections of mankind' (E293). His reasons for thinking this are as follows. If you desire something as a means to some further end, it is possible to give a reason why you want it. For example, you can give a reason why you want to take exercise, if you want to take exercise as a means to attaining health, and you can give a reason why you want health if you desire health as a means to the pleasure it brings, but you cannot give a reason why you want pleasure, for pleasure is an ultimate end, and is desired for its own sake (E293). Hume next argues 'Now as virtue is an end, and is desirable on its own account, without fee or reward, merely for the immediate satisfaction which it conveys; it is requisite that there should be some sentiment which it touches, some internal taste or feeling, or whatever you please to call it, which distinguishes moral good and evil, and which embraces the one and rejects the other' (E293–4).

Comments

This argument is another *ignoratio elenchi*. Hume is right in thinking that we do desire virtue for its own sake (though, incidentally, this does not mean that we cannot also desire it as a means to some further end, for example, having a good reputation). And from this he quite rightly thinks that it does follow that we cannot give a reason why we desire virtue—or, more accurately, that a man who desires virtue only as a means, and not at all as an end, cannot give a reason why he desires virtue. But it does not at all follow from this that virtue is *recognized* by a sentiment, or that it is a sentiment which distinguishes moral good and evil. Many men desire power for its own sake, which means that they cannot give a reason why they desire power (as opposed to a causal explanation of how it comes about that they desire power). This does not mean that, without a desire for power, men would not be able to recognize power when they came across it, or tell when they or other people possessed it.

VII

HUME'S POSITIVE CONCLUSIONS

A. *A TREATISE OF HUMAN NATURE*, BOOK III, PART I, SECTION II

HUME puts forward his own answer to the question how moral distinctions are discerned in Section II of Part I of Book III of the *Treatise*, and in the *Enquiry*, Section I and Appendix I. I shall first discuss what Hume says in the *Treatise*. As one might have anticipated, since moral distinctions are not derived from reason, or by the comparison of ideas, they must be discovered by means of our impressions, or by a moral sense. 'Morality, therefore, is more properly felt than judg'd of; tho' this feeling or sentiment is commonly so soft and gentle, that we are apt to confound it with an idea ...' (470).

The impression arising from virtue is agreeable, and that arising from vice disagreeable (470). 'No enjoyment equals the satisfaction we receive from the company of those we love and esteem; as the greatest of all punishments is to be oblig'd to pass our lives with those we hate or contemn' (470).

'To have the sense of virtue, is nothing but to *feel* a satisfaction of a particular kind from the contemplation of a character. The very *feeling* constitutes our praise or admiration. ... We do not infer a character to be virtuous, because it pleases: But in feeling that it pleases after such a particular manner, we in effect feel that it is virtuous' (471).

Hume discusses the objection that 'if virtue and vice be determin'd by pleasure and pain ... any object, whether animate or inanimate, rational or irrational, might become morally good or evil ...' (471). He replies, firstly, that the pleasure aroused in us by the contemplation of virtue is (introspectively) different in kind from other pleasures (472). Further—though Hume himself fails clearly to distinguish this point from the previous one—''Tis only when a character is considered in general, without reference to our particular interest, that it causes such a feeling or sentiment, as denominates it morally good or evil' (472). Secondly, virtue, which excites pleasure, and vice, which excites uneasiness, must necessarily be placed in ourselves or in others, and excite pride or humility in the former case, love or hatred in the latter. This distinguishes them from the pleasure or pain aroused by inanimate objects, which objects need bear no relation to people (473). (A sunset, for example, excites pleasure, but since it is not related either to ourself or to others, by, say, belonging to us or them, it cannot excite pride or humility, love or hatred.) He goes into this question in more detail in Book II of the *Treatise*.

Obviously, many different kinds of thing, perhaps an infinite number, arouse our approbation or disapprobation. Hume thinks it incredible that facts about what things arouse our approval should be irreducible facts (473). There must, he thinks, be some general principles from which such facts as that honesty, industry, and wit arouse approval are derived (473). In other words, these things must have something in common, which explains why they are all approved of. (Later it will appear that what they have in common is that they are all characteristics useful or agreeable to themselves or others.) Since virtue and vice are distinguished by the pleasure they occasion us (475), the question of the origin of the rectitude of virtue or the depravity of vice (i.e. the question what these things have in common, which explains why they are virtues or vices) becomes the simple empirical one: Why do certain qualities (the virtues) give us pleasure, and certain other qualities (the vices) give us pain? (475–6.) The rest of this section (473–5) is taken up by a discussion of the question whether the moral sentiments are natural or artificial, which it would be irrelevant to go into.

Comments

(1) Hume gives no clear account of the relation between the judgement I express when I use the words 'Laziness is a vice' and the feelings of disapproval which laziness arouse in me. Is the former just a judgement to the effect that contemplating laziness, or supposed laziness, causes me to feel disapproval? When, however, he suggests that the origin of moral beauty or depravity reduces to the question why any action or sentiment gives satisfaction, it looks as if he means 'satisfaction to mankind'. In this case judgements such as 'Laziness is a vice' should be judgements about the feelings of disapprobation laziness arouses in men generally, not just in me (475). Sometimes Hume simply fails to draw a distinction between the judgement and the feelings—'The very *feeling* constitutes our praise or admiration' (471).

(2) It seems to me that Hume exaggerates the pleasurableness of the feeling of approbation of virtue, and the painfulness of the feeling of disapprobation for vice. It is sometimes quite pleasurable to disapprove of vice in those we dislike. He compares the satisfaction we receive from the company of those we love and esteem with the pain we get from the company of those we hate and condemn, but fails to compare what pleasure we get from those we love, but do not esteem, and what pleasure we get from those we esteem, but do not love. It is true that Hume regards esteem as a kind of love (608 n.), but it is still possible to compare the pleasure we get from the company of those we love, in the form of esteeming them, with the pleasure we get from the company of those we love in the ordinary way. He loves to attribute more amiability to virtue than is wholly plausible, even on his own theory, in those cases when the advantages of certain useful, as opposed to agreeable, virtues are remote. Some virtues arouse respect, rather than love.

(3) Hume says, ''Tis only when a character is considered in general, without reference to our particular interest, that it causes such a feeling or sentiment, as denominates it morally good or evil' (472). He speaks as if this were a synthetic proposition, but it seems to me more likely that it is analytic; we do not *call* that favourable attitude we have to what is merely to our interest, or to the interests of our friends or relations, moral approval. Moral approval is, *by definition*, that favourable attitude we have to a character when it is 'considered in general'. Its felt intrinsic introspectable qualities are not of much importance.

But what is it to consider a character 'in general' or, as Hume says later (475), 'upon the general view or survey'? I think Hume must mean something like 'consider what our attitude to the action or character would be, if we ourselves were not personally affected by it'. We should in this case, I think, take into account its effect on us, but not be weighted by the fact that this is an effect on *us*, but consider it simply as an effect on *someone*. We will not, I suppose, actually *have* the attitude which we would have if our interests were not affected, but we can estimate what this attitude would be, and judge the character in question to be a virtue or a vice accordingly.

(4) Hume's second answer to the difficulty—that a virtue cannot be whatever excites pleasure or a vice whatever excites uneasiness, for in this case inanimate objects could be virtuous or vicious— involves a consideration, which must necessarily be a brief one,

of what he says about pride and humility, love and hatred, in Book II of the *Treatise*. There he states that love and pride, hatred and humility are aroused by qualities which produce pleasure, in the case of love and pride, or pain, in the case of hatred and humility, provided these agreeable or disagreeable qualities belong to or are related to ourselves, or to those to whom we are in some way or other connected. Hence they cannot be directed towards inanimate objects, though they can be directed towards the possessors of inanimate objects—the possessor of a beautiful house, for example—if these objects are agreeable or the reverse. Approval and disapproval are like love and hatred, pride and humility—indeed, Hume sometimes says that they are a kind of love or hatred—in that they are aroused by qualities which independently produce pleasure, in the case of love and pride, or pain, in the case of hatred and humility. However, the relation to oneself—you can, for example, be proud of something only if you possess it, or if it is possessed by someone related to you, your father, say—which is necessary in the case of pride, is not necessary in the case of approval or disapproval. One can approve of any man, however little related he is to oneself. Furthermore, approval and disapproval must be impartial in a way in which love and hatred do not have to be. It is not wrong to love those who are specially related to oneself—one's parents, for example—more than those who are not, but it is wrong to approve of those who are specially related to oneself more than those who are not. Hence we must make allowances, as Hume himself says, for the ways in which our feelings of approval are affected by such special relationships, whereas there is no need to make such allowances in the case of love and hatred. No correction need be made to allow for the fact that absence makes the heart grow fonder, but correction must be made to allow for the fact that contiguity makes approval stronger.

It is simply a mistake to suppose that we can only be proud of people (ourselves). We can be proud of or ashamed of (there is no such expression as 'be humble of') our possessions. And it is probably a mistake to suppose that we cannot love or hate inanimate objects.

(5) Hume's claim that the virtues and vices must be deduced from some original principle has always seemed to me to be extremely plausible. It is tantamount, of course, to a rejection of that moral theory known, many years after Hume wrote the *Treatise*, as Intuitionism. According to Intuitionism, truth-telling, debt-

paying, promise-keeping, etc. are right simply because they *are* truth-telling, debt-paying, etc., and not because they are practices conducive to the welfare of the community which has them, nor for any other reason. Hence there are as many rules as there are classes of action which are obligatory, and any attempt to derive these rules from one single ultimate principle is misguided, and doomed to failure. I personally find the amount of unsystematized anarchy which such a theory tolerates offensive, and the too easy acceptance of the *prima facie* appearance that there is this amount of disorder among our moral principles defeatist.

B. HUME'S POSITIVE VIEW IN *AN ENQUIRY CONCERNING THE PRINCIPLES OF MORALS*

Hume's positive view in the *Enquiry* does not differ greatly from that in the *Treatise*. What difference there is consists in the fact that he settles fairly firmly for the view that a virtue is any quality or character which arouses approbation in the human mind (not just in the person judging that this quality is a virtue), and that which qualities are virtues and which are vices is settled by an empirical enquiry into which qualities arouse approval, and which disapproval, in men (E289). But this simple picture does not by any means fully accord with everything Hume says everywhere, and especially with what he says in the *Treatise*. In Section D of this chapter I shall make an attempt to unravel some of the numerous, and, unfortunately, inconsistent, strands in Hume's epistemological views.

C. THE CAUSES OF APPROVAL AND DISAPPROVAL

We have seen that Hume thinks that 'to have the sense of virtue, is nothing but to *feel* a satisfaction of a particular kind from the contemplation of a character' (471). Hume says later that virtues are those characteristics which are useful or agreeable to ourselves or others (E216). This raises the question: 'Why do characteristics, which are useful or agreeable to ourselves or others, cause us to feel pleasure?' (E213.) (Notice that it is improper to ask why virtue gives pleasure. That virtue gives pleasure (of a special kind) is an analytic proposition. What does not give pleasure is not virtue.)

According to Hume, those characteristics which are virtues give us pleasure because we sympathize with their usefulness or agreeableness to those people to whom they *are* useful or agreeable. It is pleasant to sympathize with others' pleasures, and painful to sympathize with others' pains. Hume has two different accounts of sympathy, the former of which is more prominent in the *Treatise*, the latter more prominent in the *Enquiry* (E212–32). The first account is as follows. Sympathy is just a desire for the happiness of others, and the pleasure we get from their happiness is just the pleasure of having a desire satisfied.

His second account is more complicated (317–24). Put very briefly, it is that sympathy is just a tendency to have those feelings and beliefs which other people have (317). Thus we feel sympathetic pain when we tread on another's gouty foot, and sympathetic pleasure from the enjoyment another derives from his wife, his family, his friends, and his possessions. We feel, sympathetically, emotions not only in consequence of the belief that these are felt by another, but also from the thought, unaccompanied by belief, that these are felt by another. For example, we feel sympathetic fear as a result of seeing an actor play the part of a man in terror on the stage, though we do not actually believe that he feels terror (E221–2). We may feel sympathetic pleasure from viewing a fertile landscape, although it is deserted, and there are no people who reap actual benefits from it, with which benefits we may be said to sympathize (E228 n.).

In more detail, Hume's account is as follows (317–24). Seeing someone being injured (a cause of pain) or groaning (an effect of being in pain) brings to mind an idea of pain, which idea, of course, is a faint copy of an actual impression of pain (317). This idea of pain is converted into an impression of pain, because it is related to my impression of myself by such relations as resemblance, contiguity, and causation (317). Hence, either believing that someone is in pain, or thinking of someone (for example, an actor on the stage) as being in pain, causes me to feel pain, but in a fainter degree.

The relation of resemblance enters into the picture because all men are roughly similar, and there are no emotions which any man experiences which another man is not capable of experiencing (318). Contiguity comes into the picture because we sympathize more with opinions and emotions in those who are in our physical neighbourhood than we do with those of people who are not (318).

Since our impression of ourselves is at least as lively as any other, and is always with us, it confers its force and vivacity upon our ideas of others' beliefs and emotions by means of the above-mentioned relations (317). (These ideas will be forceful and lively ideas if we believe others have the beliefs and emotions they seem to have; otherwise they will not be. When we believe they have the emotions they seem to have, this belief is based on a species of causal argument, from an impression of the overt signs of their emotions, to a forceful and vivacious idea of the emotions themselves (320).)

This is an oversimplified account of Hume's views on sympathy. Hume's account of sympathy has ramifications throughout the *Treatise*, and to discuss all of them would take too long. In any case, the subject is peripheral to an understanding of Hume's moral epistemology. A longer treatment may be found in *Sympathy and Ethics* (1972) by Philip Mercer, and in *Passion and Value in Hume's Treatise* (1966) by Páll S. Árdal.

Comments

(1) Hume's first account of sympathy is a little puzzling, in that in the *Treatise* he states that there is no such thing as benevolence, in the sense of a desire for the welfare of people as such, regardless of who they are, or how they are related to us (481–2).

(2) If the pleasantness of approval were just the pleasantness of satisfying our benevolent desire, it would be very difficult to explain why it is not just one satisfaction among others. It is fairly obvious that it is *not* just one satisfaction among others, for we do not feel that we ought to act rightly if it so happens that we want the pleasure of satisfying our benevolent impulses, but that, if we prefer some other kind of pleasure, we need not.

(3) Hume's first account of sympathy makes it very difficult, therefore, to explain why we feel guilt if we do not act rightly. If moral approval is just the pleasure we get from satisfying one desire among others, there should be no reason why we should feel guilty if we choose to satisfy some other desire instead.

(4) Hume's second account of sympathy is much more difficult. It has two stages. The first stage explains how we get the idea of another person's having a certain feeling. The second stage explains how this idea is converted into an actual impression of this feeling, so that we actually feel, though less intensely, the emotions which we think of others as having.

The first stage is straightforward, as it is, according to Hume, a normal causal inference. At least, it is a causal inference when we actually believe that someone has the feeling with which we sympathize. When we do not believe this, as when we sympathize with the feelings expressed by an actor, poet, or painter, I think that Hume should have said that we have a tendency to make the causal inference, which tendency is inhibited by our knowledge that the circumstances, in which the signs of someone's having these feelings are displayed, are unusual.

The second stage is less straightforward. Hume pretends to think that the present impression (the impression of ourself), and the relations of causation, resemblance, contiguity, and so on, are playing their usual role (see pp. 40–6, above) but this is not so. Their usual role is to enable an impression of one thing to produce a forceful and vivacious idea of another thing, or to turn an idea into a forceful and vivacious idea. Sympathy, however, involves an idea, whether forceful and vivacious or not, being converted into an impression. In other words, causation, contiguity, and resemblance usually produce belief, but, where sympathy is concerned, they do not produce belief that someone has a sentiment (this belief, if it exists at all, precedes my sympathy) but the actual sentiment.

For this reason, the function of the present impression (of myself) in sympathy is obscure. In the genesis of belief, a present impression is necessary to serve, roughly speaking, as a premiss. I need an impression of smoke, in order to have a forceful and vivacious idea of fire, because seeing the smoke is my grounds for believing that there is fire. But since what I actually, though only sympathetically feel, is not a belief, it is difficult to see why any impression is necessary, over and above those impressions which give rise to the belief that someone else is having the experience with which I sympathize.

Hume's erroneous account of propositions may facilitate his acceptance of his account of sympathy. If the proposition that Smith is in pain were just a faint image of pain, and my belief that Smith is in pain a more forceful and vivacious image of pain, it would seem easy to see how this proposition, which is just an image, can, if it becomes forceful and vivacious enough, become an impression of pain, and so become pain. But my belief in the proposition that Smith is in pain is not the same thing as a faint copy of pain, and however forceful and vivacious it gets (i.e. however firmly I believe it), it remains a belief that Smith is in pain, and never becomes an actual feeling of pain.

In sympathy, too, causation, contiguity, and resemblance are working in the reverse direction from that in which they work when they produce belief. In the latter case, they enable an impression to produce an idea; in the former, they enable an idea to turn into an impression. Hence constant conjunction (causation), which occupies a paramount position in Hume's account of the genesis of belief, dwindles into near insignificance in his account of the transformation of an idea into an impression. This, presumably, is because, though it is quite easy to see why, if two impressions have been constantly conjoined in the past, the recurrence of one of them should produce a lively idea of the other, it is very far from obvious why, if two impressions have been constantly conjoined in the past, an *idea* of one should *turn into* an impression, not of the other, but of the one of which it is itself a copy. The only reason that I can see why contiguity and resemblance should produce belief is that they are often conjoined with constant conjunction, so to speak. (Chimney-pots are not only contiguous with grates, they are also constantly conjoined with them. Contiguity alone would not justify me in inferring an unobserved grate from an observed chimney-pot.) But, in the case when an idea is transformed into an impression, there cannot be this reason for expecting contiguity and resemblance to be conjoined with constant conjunction, for constant conjunction does not explain how an idea gets *transformed into* an impression.

We have seen that the impression of ourselves cannot enter into sympathy as a premiss. In any case, since it is always present, it could not be a premiss from which anything can be inferred, for it would be constantly conjoined with all impressions equally. It enters into Hume's account of sympathy, in that, so far as I can see, it mixes with my idea of pain, and so becomes an idea of my being in pain, which then becomes an impression, because I have an impression of myself, and whatever is mixed with it becomes an impression. All that I can say is that I find this virtually incomprehensible.

(5) Hume has maintained elsewhere in the *Treatise* (633–6) that there is no impression of the self.

(6) In Hume's account of belief, resemblance works in producing, from an impression, an idea which resembles it. The same is not true of the manner in which resemblance works in sympathy. My idea of someone's sorrow, it must be conceded, is turned into an impression of sorrow, which, of course, resembles the idea. This, however, does not explain *why* ideas produce sympathetic impressions; if the relation of resemblance did not exist between the idea and the impression, the impression simply would not *be* a case of sympathy. Hume speaks of all men having roughly similar natures. This must mean that all men have similar dispositions to feel, or that they all have actually similar feelings. In the latter

event there would be no need for sympathy, for I would already have all the passions which sympathy is supposed by Hume to produce. In the former event, there is no resemblance between impressions or ideas, for a disposition to have impressions is neither an impression nor an idea.

(7) The relations which Hume mentions in talking about sympathy, contiguity, and resemblance, he describes as holding between people. When he is talking about belief, these relations are usually supposed to hold between impressions and ideas.

(8) Hume can give no explanation of the fact that, though I may feel pleasure or pain when others are in pleasure or in pain, I do not see sympathetic views when others see views, nor do I hear sympathetic noises. Nevertheless, if Hume's theory explains the former, we ought also to have the latter.

(9) There is no doubt that sympathy, in Hume's sense, exists. We do sometimes feel, in sympathy, the emotions and share the beliefs which we believe other people have, or think of them as having. And sometimes, too, the contiguity of the person who has the emotions or beliefs increases our sympathy, and we may often sympathize more with people like ourselves. Sometimes, even, we sympathize more with relatives than with other people. The explanation why we feel more sympathy for some people, some emotions, and some beliefs, however, must be more complex than Hume allows. If I witness a bout between two wrestlers, I may sympathize both with the triumph felt by one and the humiliation experienced by the other, but I am more likely to sympathize with the feelings of one of them only. The reason for this cannot be contiguity, because both wrestlers are equally contiguous. Neither of them is likely to be my relation. It would need a lot more arguing than Hume gives it to show that I sympathize most with the one I most resemble. I may sympathize most with the one who comes from my town, county, or country, but this is not a relation Hume mentions, and it is not always operative. I may also sympathize most with the one with whom I most closely identify myself, but Hume can give no account of this.

(10) As with Hume's first account of the genesis of moral approval, Hume's second account leaves out what is specifically moral about moral approval. A tyrant lays waste towns and villages in order to increase his dominions. I may feel sympathetic pain with the victims, or sympathetic pleasure with the tyrant, or both, but what is there in this to dictate that I should give any moral preference to the former, and condemn the tyrant's actions? The

pain suffered by the victims greatly outweighs the tyrant's pleasure, but it does not follow from this that I do not get more sympathetic pleasure from contemplating the tyrant's pleasure than I do from contemplating the victim's pains. Hume holds that if I do get more sympathetic pleasure than pain from a situation like this, this must be, say, because I am specially related to the tyrant, and that I should then allow for this. In this case, I should pronounce something wrong not because of the sympathy I do feel, which is for the tyrant, but because of the sympathy I believe I would feel, if I were impartial. Nevertheless, that I would feel sympathetic pain for the victims, were I impartial, seems to leave out the moral element in my approval. To it must be added, at the very least, a hostility to the action, and others of its sort, which is distinguished from other forms of hostility by being based on the thought that it or they are not the kind of action society ought to tolerate.

(11) If Hume's account of the genesis of our moral feelings were correct, our moral feelings would be a great deal milder, more humane, and more felicific than they in fact are. Hume's account is intended to explain, and can only explain, why we approve of characteristics which are useful or agreeable to ourselves or others. But in fact we sometimes approve of characteristics which are not useful or agreeable to ourselves or others. As much misery has been produced in the world by wrong-headed people acting in ways they supposed to be virtuous as has been produced by people acting in ways agreed by everyone, including themselves, to be vicious. Many martial and many monkish virtues have been approved, in spite of their being known to bring happiness to no one, and misery to many. Hume, it is true, condemns the monkish virtues, but this is not the point. Hume may well be right in saying (E270) that men are quite wrong to regard as virtues characteristics which are useful or agreeable neither to their possessor nor to anybody else, but the question is 'How can he even explain how it is that they have ever been regarded as being virtues, if approval simply is that sympathetic pleasure we get from contemplating the usefulness or agreeableness of certain characteristics?'

(12) We sympathize with pleasures and pains that we believe people have, whether these beliefs are true or not. We do not sympathize with pleasures and pains people have, if we do not know that they have them. Hence, Hume should not have said that a virtue is a characteristic of which, on account of sympathy, we approve. He should have said that it is a characteristic of which

we would feel approval if we were aware of the pleasures to which it gave rise, and also had no false beliefs about these pleasures. The monkish virtues, for example, are not virtues, although some people may feel sympathetic approval of them because they falsely believe that having these virtues brings pleasure to their possessors in an after life.

D. FIVE VIEWS THAT MIGHT HAVE BEEN HUME'S

Hume was not quite clear about what his own alternative to rationalism was. Hence it is necessary to say that one or other of the views to be discussed in this section might have been his. Some of these views fit in with some of the things he says, others with other of the things he says. None of them fits in with everything he says.

(1) Moral judgements are about the judger's feelings

Hume sometimes writes as if he held the crude view that to say that an action was wrong or vicious was to make a statement about the feelings contemplating that action aroused in *the person making this judgement* (469). However, his more considered view was often that a moral judgement was a judgement not about the actual feelings which contemplating the action being morally assessed aroused in the person making the judgement, but that it was one about the feelings he would have, if certain allowances were made for such facts as (a) he tended to disapprove more strongly of actions near in space and time than he did of actions remote in space and time (581), (b) that he tended to disapprove more strongly of actions he witnessed than those he did not witness (582), and (c) that he tended to disapprove more strongly of actions performed against him by enemies than he did of similar actions performed on his behalf by friends (583). He also suggests, though he does not explicitly assert, that such judgements would have to be corrected to allow for the fact that sometimes the beliefs, about the action, which caused him to feel approval or disapproval of it, might be mistaken or incomplete (E173). Hume seemed to suppose that, were everybody to make such corrections, there would be a great measure of agreement between the moral judgements of different people, just as, when everybody allows for such facts as that distant objects look smaller than near ones, there is a high level of agreement between different people about the perceptual properties of things.

This view of Hume's fits in ill, as we have seen, with his contention that morality alone, unlike beliefs, can move us to action, for beliefs about my feelings of approval are beliefs, like any others, and move us to action as little as other beliefs do. (Beliefs about my feelings, however, are specially privileged, when compared with other beliefs, in that, though they do not alone move me to action, I must, when they are true beliefs, have the feelings I truly believe I have, and these feelings *may* alone move me to actions. This remark draws attention to a mitigating circumstance only for the crude view that moral judgements are about my actual beliefs. The feelings I do not have, but would have, if certain allowances were made for the fact that my actual feelings may be biased or rest on mistake, obviously cannot move me to action.) This view, too, also fits in very ill with Hume's contention that morality does not consist in any matter of fact, for that I approve of something is a matter of fact. One might, of course, say (rightly) that though the fact that I approve of something is a matter of fact about the action, my approval is nevertheless in me (469), and not in the action, and a change in my feelings does not imply that the action itself has changed, though it does imply that I have changed. This is true, but, nevertheless, there is no earthly reason why I should alone be moved to action only by beliefs which are not about an action's intrinsic properties; if one will not alone move me to action, the other will not either. However, with a special sub-class of beliefs about the non-intrinsic properties of actions, that is, beliefs about how these actions affect the feelings of the person morally judging them, the case is slightly different. As we have seen, if the belief in question is that I am not personally indifferent to something, then, if this belief is true, I will not be indifferent to this thing—though, of course, it still is not the belief, but the feelings I truly believe I have, that move me to action. Hence, even if moral beliefs are beliefs about my own actual feelings, they still do not alone move me to action.

If Hume had supposed that moral judgements were about my actual feelings—the crude view mentioned at the beginning of this section—reasoning, even inductive reasoning, would be unnecessary to establish these moral judgements, for I do not normally infer that I approve of something; I know this by introspection. However, if we consider what I have described as Hume's more considered view, inductive reasoning is needed to make the corrections to our feelings which Hume thought necessary. Just as I know inductively that, from a distance, mountains do not look the colour they usually look, and so refuse to say they are the colour they look from that distance, so I may realize on inductive grounds that I do not feel the approval which I usually feel of generosity when this generosity is manifested by an enemy, and so refuse to say on that account that my enemy's generosity is not admirable. We must, however, concede to Hume that on neither the crude nor the modified view would it be possible to demonstrate *a priori* the difference between right and wrong, and so concede that, if either is true, moral distinctions are not apprehended by deductive reasoning. The crude view fits in extremely well with those passages in which Hume speaks as if we discover the morality of an action by looking within our own breast (469), and finding there a feeling of approval, in which case the morality of an action is not the conclusion of a piece of reasoning, and is not inferred. The more considered view, however, makes it impossible to discover what is right simply by looking within our own breast on only one occasion. We must arrive inductively by probable reasoning at rules about what actions arouse feelings of approval in us, and in what circumstances, in order to correct our actual feelings, on the grounds that the circumstances which produce them are unusual. If Hume thought that moral judgements are about the judger's feelings, it is difficult to say why he thought that an 'ought' could not be derived from an 'is', for factual descriptions of the judger's feelings would, on such a theory, entail moral judgements to the effect that the actions he disapproved of were wrong.

Hume's crude view, that moral judgements are judgements about the feelings aroused by an action in the person making the judgement, is not even remotely plausible. The main, though not the only, reasons for which it must be rejected are these. (a) It implies that, if one person says that an action is right, and another person says that the *same* action (as opposed to the same kind of action) is wrong, they can both be right (because their feelings can differ). But obviously, no one person can say to another: 'When you say that Winston Churchill was *wrong* to order the destruction of the French fleet at Oran, I quite agree that what you say is true, though I think he was *right* to order its destruction.' (b) It implies that one and the same action (as opposed to classes of actions) can be truly judged to be right at one time, and truly judged to be wrong at a later time (because people's feelings can change). No one, however, can say such things as 'Once upon a time I thought Winston Churchill was right to order the destruction of the French fleet. Now I think he was wrong, but, since

I was right when I thought he was right, there is no need for me to change my mind or reject my former judgement.' If someone says at t_0 that something is right, and at t_1 that this same thing is wrong, one or other of his two judgements must be mistaken, however much his feelings may have changed in the interval. (c) It implies that, if I say that something which I do in fact regard with approbation is right, I judge truly that it is right, even if I would not have felt this approbation of it had I not been mistaken or ignorant of matters of fact about it.

Hume's more considered view need not fall to the last (E290) of these three objections, and does not so fall to the extent that he allowed for the fact that corrections to the judger's actual sentiments must be made to accommodate the fact that he might be mistaken or ignorant about matters of fact concerning the action judged. Any view, however, according to which moral judgements have a covert reference to the person making the judgement, to *him*, or *his* feelings, or the feelings which *he* would have, must fall to the first two of the above objections, for then different judgers are making statements about different people, and their judgements need not be logically interrelated in any way.

(2) Moral judgements are about the feelings of mankind

Hume's more considered view naturally leads to the doctrine (attributed by Broad to Hume in *Five Types of Ethical Theory*, 1930), which *sometimes* was Hume's, that moral judgements are not judgements about the speaker's actual or corrected feelings, but about the feelings which any man has (E289), or would have, if something like the above-mentioned corrections were made to his actual feelings. On such a view, moral judgements or beliefs are still beliefs about a matter of fact, though, again, one might say that, though the fact that an action is generally approved of is a fact about the action, the approval lies in the people observing the actions judged of, not in the actions themselves, and that a change in these feelings is not a genuine change in the action. Since moral judgements are still beliefs, it is still impossible for Hume to explain how they can alone move us to action, and it is in any case obvious that a belief about what actions are generally approved of will leave us indifferent unless we care about what is generally approved of, or have in us some passion to do what men generally approve of. Although one can be moved to action by one's own approval of an action (though not by the belief, even the true belief, that one approves of it), one not only cannot alone be moved to action by the belief that others approve of something; one also cannot be moved to action by other people's feelings of approval, though, of course, they must have these feelings if my belief that they have them is true. On this view, too, reasoning, in the form of inductive reasoning, is necessary in order for me to know what is right and wrong, for it is necessary in order for me to know what men approve or disapprove of. Presumably I know this by induction by simple enumeration; Tom, Dick, and Harry, etc. approve of this sort of thing, and no one does not approve of it, so all men do. On the other hand, to the extent that Hume held this view, he is absolutely right in thinking that no amount of *a priori* deductive argumentation will tell us what is right and wrong, for no amount of such argumentation will tell us what men actually do approve of. This theory, too, is very difficult to square with Hume's contention that it is impossible to derive an 'ought' from an 'is', for, if it is true, moral judgements will follow from (because they are logically equivalent to) judgements about how men in general feel about actions.

To the extent that he holds the views we have been discussing in this and the preceding section, Hume is getting close to holding an 'ideal observer' theory of moral judgements. Such a view was first suggested by the great economist Adam Smith, who, in *The Theory of Moral Sentiments*, first published in 1759 put forward the view that moral judgements were not judgements about the feelings of any actual person or group of people, but about the feelings of an impartial spectator. To the extent that Hume thought that we had to make allowances for the bias of actual observers, he was moving towards an ideal observer theory. Again, beliefs about how an ideal observer or impartial spectator would feel cannot move us to action without the co-operation of passion; the fact that an impartial spectator would approve of an action is a matter of fact about it, though an extrinsic fact about it; reason, albeit inductive reason, would be necessary to establish what an impartial spectator would feel; and, on an ideal-observer theory, it would be possible to deduce an 'ought' from an 'is', for moral judgements would follow from factual judgements about how impartial spectators would feel.

The view that moral judgements are about the feelings (corrected or otherwise) which contemplating actions arouse in men should be rejected, on the grounds that it is logically possible to believe without logical inconsistency that all men do or would approve of something which you personally think is wrong, or vice versa. There is not even any inconsistency in thinking that an impartial spectator might feel approval of something which you personally regard as wrong. The only way of defining an impartial spectator to avoid this possibility is to say that an impartial spectator is a spectator who always approves of what is right, and disapproves of what is wrong, but such a definition immediately makes the theory, that a right action is one which would be approved of by an impartial spectator, circular. Furthermore, it is not really very plausible to suggest that, in order to make up our minds about what is right and wrong, it is necessary to conduct an empirical survey into the feelings aroused by the actions we pronounce to be right or wrong in mankind as a whole.

(3) A moral sense theory

Hume describes himself as holding a *moral sense* theory (470), but not much importance need be attached to this, as he is inclined to speak (wrongly) of 'sense or sentiment' as if it mattered little which of the two words he used. There is an important difference between sense and sentiment, however. My senses give me information about the things I see or hear or feel. Sentiments (like anger or fear or approval), are, however, emotional reactions to beliefs (for example, the belief that the animal coming towards me down the street is a tigress or a woman) and do not give me information about the things which, because of the beliefs I have about them, arouse in me the emotions they do.

A moral sense theory would accord ill with Hume's view that there are no moral matters of fact, for one would have supposed that, if there were a moral sense, its function would be precisely to discover such facts about an action as that it was right or wrong. Indeed, one would have supposed that it would discover to us properties of the action itself, not simply extrinsic facts about actions such as that people reacted to them in various different ways. It looks, too, as though, if a moral sense enabled us to arrive at true beliefs about actions, these beliefs would alone move us to action as little as any other belief. Something can be said on Hume's behalf in reply to this difficulty, however. If accidentally I put my hand in nearly boiling water, and the water feels painfully hot, my feeling of pain both reveals to me something about the water, that it is hot, and provides me with an incentive to move my hand. Perhaps Hume thought that my moral 'sense or sentiments' were similar to my feeling of pain, and both revealed to me properties of the action which aroused the sentiments, that

it was right or wrong, and provided me with an incentive to perform or avoid such actions. If this is what Hume thought, it is
not, strictly speaking, my *belief* that an action is wrong which,
without the co-operation of passion, moves me to action, but the
sense-cum-sentiment which reveals to me that it is wrong which
moves me to action.

To the extent that Hume held a moral sense theory, he is very
easily able to explain how my moral beliefs are not arrived at by
a process of reasoning, whether deductive or inductive reasoning.
It would seem natural to say that I just see that something is red,
and do not infer that it is; hence, if it is my moral sense that tells
me that an action is wrong, I just sense that it is wrong, and do
not infer that it is. A moral sense theory, too, would easily enable
Hume to explain how it is that an 'ought' cannot be deduced from
an 'is'. Just as my senses tell me that something is yellow, and
the fact that it is yellow cannot be deduced from any other fact
about it, but *has* to be revealed to me by my sense of sight, so,
if my moral sense tells me what actions are right and wrong, that
these actions are right or wrong is something which cannot be
arrived at independently of my moral sense; it cannot be deduced
from any other facts I know about the actions in question, and
has to be revealed by my moral sense. Indeed, a moral sense theor-
ist might argue with plausibility that, just as my sense of sight
reveals to me such indefinable characteristics as redness, so my
moral sense reveals to me the indefinable characteristic 'wrong-
ness'. If a characteristic is indefinable, its presence cannot be
deduced from the presence of the characteristics in terms of which
it is defined, because there are no such characteristics.

Hume, however, when he is not explicitly describing his own
theory as a moral sense theory, almost always speaks as if all that
is involved in our apprehension of morality is emotional reactions
to beliefs, which reactions, since I ought properly to have complete
knowledge of the action I am assessing morally *before* I pronounce
it to be virtuous or vicious, can give me no new knowledge of the
actions which arouse the feelings (E290). One of the clearest cases,
so far as I know, in which he does speak in the manner appropriate
to one who holds a moral sense theory occurs on page 471 of the
Treatise: 'We do not infer a character to be virtuous, because it
pleases: But in feeling that it pleases after such a particular manner, we in effect feel that it is virtuous' (471). Here he seems to
speak as if we feel a character to be virtuous, much as we feel that
the penny in our pocket is hard.

A moral sense theory is no more plausible than the other theories, elements of which we have said are to be found in Hume. There is no such thing as a moral sense organ. There are no 'moral light waves' emanating from the actions which we morally assess, and we do not need to take up any special spatial position in relation to these actions in order to sense their moral qualities, as we do in order to perceive the properties of material objects. We can also pronounce upon the morality of actions remote in time as easily as we can upon the morality of contemporaneous actions. We make our moral assessment of actions in virtue of the other characteristics we believe them to possess, and our assessments are liable to be mistaken to the extent that the beliefs upon which they are based are inaccurate or incomplete. We do not, however, pronounce a balloon to be red in virtue of its other characteristics, but 'straight off', so to speak, and there is no need to revise our opinion that it is red when we discover that we are mistaken about it in other ways. A moral sense theory must also be rejected because it can give no account of the fact that, if an action has some moral quality, then any other action, similar to this action in all respects *save* that of possessing this moral quality, must *also* resemble it in possessing the moral quality. In other words, if two things are to be morally assessed differently, they must differ from one another in some (other) way, but there is no reason at all why this should be true of *sensible* characteristics such as red. That one balloon is red and another green could (logically) be the only difference between them.

(4) A non-propositional theory

A theory which would fit in very well with many of Hume's views and arguments would be the theory which holds that so-called moral judgements are not really judgements at all: the function of sentences containing moral words is not to enable us to make statements, which will be true if the facts are as we say them to be, and false otherwise, but some other function, for example, the function of enabling us to issue commands to other people ('Speak no evil'), to express our emotions (like 'Alas that so many people behave like that'), to express a wish ('Would that men behaved like that more often'), and so on. (There is, in fact, nearly as much disagreement, among the philosophers who think that the function of ethical language is not to state truth, about exactly what non-propositional function ethical sentences have, as formerly there was among those philosophers who thought the function of ethical sentences *was* to state truths, about just what kind of truth they stated.)

Non-propositional ethical theories fit in well with Hume's views because, since they maintain that the function of ethical sentences is not to state truths, but, say, to express imperatives, it follows that there are no ethical truths which they state, and so no ethical truths to be stated. Consequently, if this view is acceptable, it follows that Hume is right in thinking that morality does not consist in matters of fact. What follows, of course, is not only that morality does not consist in matters of fact intrinsic to the actions morally assessed; it does not consist in any matters of fact at all, whether intrinsic or extrinsic to the actions, and so not even in facts such as that people take up certain attitudes of approval or disapproval towards them. Non-propositional theories, too, make it quite easy to explain why an 'ought' cannot be derived from an 'is'; no amount of knowledge of matters of fact logically ties one down to taking up and expressing one attitude to things rather than another, or commanding one thing rather than another. (It is commonly said that no imperative sentence can follow from any set of indicative sentences, and, though this may not be true, it is at least plausible. It is certainly true, however, that there are sets of indicative sentences which make certain imperative sentences *infelicitous*; for example, there is something wrong with commanding someone to shut the door if it is shut already.) It would also look as if the truth of a non-propositional view would justify Hume's claim that moral distinctions are not apprehended by reason, for, if there are no moral truths, there can be no moral truths for reason to discover. The situation, however, is not quite as straightforward as this; sentences which express advice, for example, do not state truths which reason might discover; nevertheless, the advice they enable us to give may be good advice or bad advice, and it would be foolish to assume that inductive reason, at any rate, played no part in determining which of these it was. It would seem natural to say that advice was good advice if the inductively established proposition that acting on it would achieve our ends quickly, reliably, efficiently, and cheaply was true. If a non-propositional theory is true, then there can be no moral beliefs to move us to action, either alone, or with the co-operation of some passion. There are, however, certain non-propositional utterances which are such that a condition of their being used appropriately is that the person using them should have certain feelings. A man using the word 'Alas', for example, is not using it appropriately if he does not feel sad. Hence certain non-propositional utterances are at any rate more closely linked to the feelings which move us to action than are beliefs, which, if suitably worded, can appropriately (though not necessarily truly) be expressed by people with any kind of attitude, favourable or unfavourable. Hence, perhaps, someone sincerely saying that a class of actions is wrong must, on a non-propositional theory, have a hostile attitude to such actions, and so must have a motive for not performing them himself, or for persuading others not to perform them.

Though non-propositional theories fit in very well with many of the things Hume says, they fit in very badly with others. They would make nonsense of holding a moral sense theory, according to which there are moral judgements, which do assert that actions have certain moral properties which our moral sense reveals to us. They would make equal nonsense of Hume's tendency to think (sometimes) that moral judgements are about the feelings actions arouse in the person making the judgement, and his tendency to hold (at other times) that they are judgements about the feelings which men in general have about the actions morally judged of.

In any case, as we have seen (p. 13), the view that the function of moral sentences is not to enable us to express statements, which may be true or false, is fraught with difficulty. Indeed, it is not easy to see how someone examining the behaviour of ethical sentences on its own merits, and not trying to strait-jacket them to conform to the demands of some preconceived epistemological theory, could have come to hold it. (Historically, non-propositional theories arose from an attempt to reject naturalism without rejecting empiricism. It was considered implausible to maintain that ethical sentences just expressed empirically verifiable propositions about the natural world, but unempirical—as well as implausible—to maintain that ethical sentences attributed non-natural properties to things. This dilemma would disappear if it could be held that, since ethical sentences were not used by us to express propositions, they did not express propositions attributing either natural or non-natural properties to things.)

We do in fact use 'true' in conjunction with ethical sentences, for example, 'It is true that I ought not to have behaved as I did, but I was in a disturbed state at the time.' Indeed, just as it is true that the cat is on the mat if and only if the cat *is* on the mat, so it is true that promises ought to be kept if and only if promises *ought* to be kept. The words 'believe', 'disbelieve', 'deny', 'wonder whether', 'suspect', 'come to the conclusion that', and many others, can precede ethical sentences just as happily as they can precede factual sentences. Ethical sentences can be used to express the antecedents or consequents of hypothetical propositions, the conjuncts in conjunctive propositions, and the disjuncts in disjunctive propositions. They can state the premisses or conclusions of arguments. What they express can be assumed for the sake of argument, overlooked, or obstinately disbelieved. As I have said, since it is customary to explain what sentences expressing a proposition are by saying that they are just those sentences with which the things just mentioned can be done, it *follows* that they express propositions. To say that ordinary people treat moral judgements as if they were propositions because they think mistakenly that they are propositions would be rather like saying that we walk because we think we have legs. It is worth mentioning that we both talk and naturally think of discovering or finding what is the right thing to do (not the same thing, *pace* many modern moral philosophers, as the deciding what *to* do, for we may decide to do something which we do not think *is* the right thing to do). This strongly suggests, what I think is true, that we regard the right thing to do not as something we decide, but something we discover—or, sometimes, fail to discover—and that what *is* the right thing to do is something, alas, independent of our wants, and of our feelings and opinions. The view, commonly held nowadays, that morality is a matter of arbitrary fiat bears so little relation to the facts of moral experience that it is difficult to understand how it could have achieved the popularity it has.

(5) Moral judgement a species of feeling

Much of what Hume says, however, falls into none of the conventional categories we have been discussing. He asserts in the *Treatise* (471) that 'To have the sense of virtue, is nothing but to *feel* a satisfaction of a particular kind from the contemplation of a character. The very *feeling* constitutes our praise or admiration.' And in the *Enquiry* he asserts much more frequently that all the understanding does is to pronounce on matters of (non-ethical) fact, that these pronouncements are the *only judgements* we make, and that the only other thing we do when we appraise something morally is to have feelings of blame or approbation (E290). (And this, too, may be what he means when he says 'Morality, therefore, is more properly felt than judg'd of' (470).) It may be a bit myopic to take what Hume says as literally as I am about to do, but nevertheless, the implications of his remarks are worth drawing out. The view they suggest accords well with Hume's assertion that morality does not consist in any matters of fact (for, on this view, my being aware of a moral distinction consists just in my *feeling* approval or disapproval, not in my judging that I do, or making any other judgement, for that matter). It is not only *compatible* with Hume's view that beliefs alone cannot move us to action, but actually *explains why* apprehending a moral distinction does move us to action, for to apprehend a moral distinction is just to feel approval or disapproval, and so, if feelings alone, unlike beliefs, can move us to action, feelings of approval can do this. If the assertion that feelings, unlike beliefs, can alone move us to action means that beliefs cannot move us to action without the co-operation of feelings, but feelings can move us to action without the co-operation of beliefs, it is not obviously true. But if it means that feelings cannot move us to action without the co-operation of (other) feelings, it *is* obviously true. To the extent that this is Hume's view, it at least goes some way towards explaining why you cannot deduce an 'ought' from an 'is', for it is not logically possible to deduce from any factual description of an action or character, however complete, whether human beings are going to react to it by approving of it or disapproving of it, or neither.

The trouble with the view we are now considering is that it is plainly untrue that *all* we do, over and above establishing matters of fact about the actions we judge morally, is have feelings of moral antagonism or approbation to them. Having these feelings is something we can do quietly, by ourselves, but we also use moral words and sentences, which we frequently utter out loud to other people. If Hume's statement that to have a sense of virtue is nothing but to feel a satisfaction of a particular kind is taken to mean that all we do, when we morally appriase an action or character, is to have feelings of approval or disapproval, it follows that moral appraisal can have nothing to do with the making of judgements. To make a moral judgement cannot be to report the fact that the action I approve or disapprove of has some attributes, which my feelings make me aware of, for feelings in this sense do not make me aware of any attributes, and, if they did, my making the moral judgement would presumably not consist in my having the feelings, but in my pronouncing, to myself or to others, that the action has the attribute my feelings reveal it to have. Nor, on this view, can my morally appraising an action consist in my saying that I have the feeling, for appraising an action morally is supposed to consist in my having the feeling, not in my judging that I have it. If I say that I have the feeling I ought, on this view to be *saying* that I am appraising an action or character morally, rather than *appraising* an action morally, for again, to appraise an action morally is just to have the feeling. To say that I think that an action is praiseworthy should, on this view, be to say that I feel approval of it, and to say that someone else thinks that it is praiseworthy should be to judge that he feels approval of it; but this is to give an account of the factual judgement that I or someone else *thinks* that it is praiseworthy, not to give an account of the moral judgement that it *is* praiseworthy.

Perhaps Hume could have said that, though to appraise something morally is simply to feel approval or disapproval of it, to *say* that something is right is to express these feelings. (Such a view would be a combination of this fifth strand in Hume's moral epistemology with the fourth.) In this case, there would be something over and above our having feelings involved in moral assessment, but it would not consist in the making of a moral judgement, for to express one's feelings about some character or action is not the same thing as to make a moral judgement attributing some attribute to them. But, as we have seen, this view is unsatisfactory in itself, and there is no evidence that it ever occurred to Hume, over and above the fact that it could explain some, though certainly not all, of the things he said.

An interesting question arises out of the distinction between the view that to appraise an action morally is to *judge* that feelings of approval or disapproval occur (in the speaker or in mankind as a whole), and the view that to appraise an action morally is simply to *have* feelings of approval or disapproval. On the first view, anyone raising the question 'What do men in fact approve or disapprove of?' is raising a moral question, for to say that I myself or mankind in general approve or disapprove of something is just (provided that due allowance is made for such things as that we approve more strongly of characteristics in a person with whom we are familiar than we do of the same characteristics in a person unknown to us, etc.) to say that it is right or wrong. On the second view, anyone raising the question 'What do men actually approve or disapprove of?' is not raising a moral question at all, but a purely sociological one. He is raising the question, which is not a moral question, 'What do people in fact appraise morally in a favourable way, and what do they appraise morally in an unfavourable way?' When, then, Hume, in the latter part of the *Treatise*, raised the question 'What is common and peculiar to the things people in fact feel approval of?' and answered it by saying that they approved of characteristics which were useful or agreeable to themselves or others, did he think that he was raising the moral question 'What characteristics are virtues?' and answering it by saying that characteristics useful or agreeable to oneself or others were virtues, or did he think he was raising the sociological question 'What characteristics do people regard as being virtues?' I think it is quite clear that he thought he was answering the moral question, or, more accurately, that he thought that the sociological question was the moral question, or that the moral question could be answered by an empirical investigation of what things men actually approved of. It is this sort of enquiry he was thinking of when he said: 'It is full time they should attempt a like reformation in all moral disquisitions; and reject every system of ethics, however subtle or ingenious, which is not founded on fact and observation' (E175). It is, as we have seen, very difficult to reconcile such a view with Hume's insistence that no 'ought' can be derived from an 'is'.

It may be, however, that in attributing to Hume, even as only one strand in his complex of views, the doctrine that all that appraising an action morally consists in is having sentiments of approval or disapproval, I am doing him an injustice. It may be that he thought that, just as perhaps, seeing a dagger is one thing, and judging that it is a dagger is another thing, so being aware of the virtue or vice of an action or character is one thing, and judging that it is virtuous or vicious is another. Hence, when he said that to have the sense of virtue was nothing but to feel a sentiment of a particular kind from the contemplation of a character, perhaps he did not intend to imply that there was not something, over and above having the feeling, which is to *judge* that this character is virtuous or vicious. And in the *Enquiry* (291) he does say 'Nothing remains but to feel, on our part, some sentiment of blame or approbation; *whence* we *pronounce* the action criminal or virtuous.' (My italics.) But in this case, he omits to give an account of how the judgement that an action is virtuous or vicious is connected with the sentiments of blame or approbation, and we are left in exactly the same position that we were in before: is the judgement about my feelings, or about the feelings of mankind, or do the feelings make me aware of the character's virtue or vice (feel that it is virtuous or vicious), or do moral words simply do some such thing as enable me to express my feelings, or to issue imperatives, exhortations, advice, or what have you, directed to-wards causing other people, and perhaps also myself, to avoid those actions which would occasion in me disapprobation if they were to be performed?

The reader will naturally want to ask 'But which of these five theories did Hume actually hold?' It is, however, impossible to answer this question, for Hume himself never explicitly in words, nor, so far as one can tell, in his own mind, distinguished them. If he had distinguished them, he would not have been as confused as he in fact was. Even to raise this question is something of an anachronism, because he was writing at a time when it was not customary to distinguish between the various possibilities. These possibilities are not always distinguished now, even by contem-porary philosophers, who have had the opportunity to benefit by reading the work not only of Hume himself but of his many distin-guished successors. However, it is possible to give a brief outline of the predominant strands in Hume's view, though even about these it is impossible to be unequivocal. On the whole, he thought knowledge of morality consisted in knowledge of what character-istics and actions aroused approval or disapproval in men. He does not distinguish clearly between the view that knowledge of morality is knowledge of what arouses approval or disapproval in me (which I may discover by looking within my own breast) and the view that it is knowledge of what arouses approval in mankind in general (which I must discover by an anthropological survey); he probably thought (wrongly) that, since all men were made roughly alike, and what one man approved or disapproved of, all will, it would not make any difference which view you held. Nor does he distinguish clearly between holding that knowledge of morality is knowledge of what characters and actions men approve or disapprove of (in which case, morality is not a matter of fact intrinsic to the things morally appraised) and holding that a moral feeling reveals qualities of characters and actions to us, that we feel *that* they possess these attributes (in which case, morality ought to be a matter of fact intrinsic to the action, though Hume says it is not; indeed, he says it is not a matter of fact at all).

Only the last three of these views make it possible to preserve Hume's claim that an 'ought' cannot be derived from an 'is'. On none of these accounts is it possible to explain how our knowledge of morality moves us to action, for knowledge of my own feelings (unlike the feelings of which I have knowledge) and knowledge of others' feelings (*like* the feelings of which I have knowledge) do not alone move me to action any more than anything else. If, however, you regard our awareness of a moral distinction as a matter of sense-perception, and, *per impossibile*, regard these moral sense-perceptions as consisting in having emotions, then, although the judgements based on these moral sense-perceptions will not move us to action, the sense-perceptions themselves will. But, of course, having feelings in the sense of having emotions does not reveal to us anything about the actions and characters which arouse the emotions, so any attempt to explain how apprehending a moral distinction alone moves us to action, by making a sense out of a sentiment, must inevitably fall to the ground. All these five views, however, square completely with Hume's main claim that arriving at knowledge of a moral distinction is not a matter of demonstration or *a priori* argumentation, but a matter of having certain impressions (viz. feelings of approval or disapproval) and arriving at conclusions empirically from them.

Finally, since I have been throughout fairly critical of Hume's views and arguments (Hume himself would not have approved of a reverence which substituted a superficial acceptance for a thorough grasp and realistic appraisal of his opinions), may I say that I think that the Scottish Hume was the greatest of British philosophers, and one of the greatest of philosophers who have lived anywhere at any time. If my own myopic gaze and passion for precise classification and distinction have served to obscure this fact from my readers, I can only implore them, if they have not already done so, to read him for themselves.

INDEX